Mathematical Induction

A Self-Study Guide to Mathematics

Volume 3

First Edition

Jianlun Xu

Copyright© 2020 by Jianlun Xu

All rights reserved. No part of this book may be reproduced, stored in a retrieval system or transmitted in any form or by any means without the prior written permission of the author.

Printed by CreateSpace, An Amazon.com Company

ISBN: 9798629868644

Preface

This book is written for high school/college students, home school and anyone interested in self-teaching math. The goal of the book is to help you to establish a solid foundation for your advanced mathematics studies and the preparation of SAT and ACT.

This book is one of a set of books, "*A Self-Study Guide to Mathematics*". Each book covers one particular topic of high school/college mathematics. The features of this book include
- Extensive coverage of a particular math topic
- Plain language to explain math concepts
- Plenty of math proofs and explanations to tell you why
- Abundant examples to tell you how-to step by step

Someone may say "I know what I learned from math class but I don't know why and how to solve new math problems step by step." If it sounds like you, then this book is for you.

Some students feel mathematics boring because they may not be trained to think in a mathematical way during their studies. When you complete this book, you will be gradually trained to think logically through plenty of proofs and examples in detail. You will find that math likes a fun game. That reminds me of the dialogue between my daughter and I. When she was in middle school she played piano and also took part in math contests. One day she asked me "Do you think that someone good at piano will also be good at math?" "Why do you ask such question?" I wondered. "In my class a boy is good at math and competes against me in a math contest. He plays piano very well too". "That is quite possible, math is harmonious like music". I answered. Yes, math is beautiful.

The set of books, "*A Self-Study Guide to Mathematics*", is your at home math tutor. Good luck in your math study and the preparation of SAT and ACT.

I appreciate the support from my wife and daughter who make this book possible.

Jianlun Xu

Contents

1 Mathematical Induction

1.1 Introduction . . . 1
    Inductive Reasoning
    Deductive Reasoning
    Mathematical Induction
    Notation and Terminologies

1.2 The First Principle of Mathematical Induction . . . 5
    Explanation
    Extension of the First Principle
    Proof Procedure

2 Variations of Mathematical Induction

2.1 The Second Principle of Mathematical Induction . . . 36
    Explanation
    Proof Procedure

2.2 Mathematical Induction by Previous Terms . . . 44
    Explanation
    Proof Procedure

2.3 Mathematical Induction by Jumping . . . 49
    Explanation
    Proof Procedure

Mathematical Induction

1.1 Introduction

What is mathematical induction? Briefly mathematical induction is a reasoning to prove a statement or formula. There are two basic reasoning methods used in proof, **inductive reasoning** and **deductive reasoning**.

▶ Inductive Reasoning

Inductive reasoning is the reasoning that concludes a general principle for a conjecture (statement/formula) by guessing a pattern or rule based on finite number of evidences. It can be briefly described like

$$particular \rightarrow general$$

Thus we say that induction gives us potential opportunity to explorer new rule or theorem for a conjecture. Let's see an example of a statement like

"All matters will expand when heated and contract when cooled."

We know the statement is true to many matters, To prove this statement is true for all matters, we check every matter in nature, such as wood, rock, metal, and so on. We work hard on it one by one and day after day and it sounds good for everything checked meets the statement until we check water that acts in different way. Now we can say that the statement is not true for all matters in nature. Next we give a mathematical example as below.

n	S_n
1	$5 = 2 \cdot 1^2 + 3 \cdot 1$
2	$5 + 8 = 2 \cdot 2^2 + 3 \cdot 2$
3	$5 + 8 + 12 = 2 \cdot 2^3 + 3 \cdot 3$
...	...
n	$5 + 8 + 12 + 19 + \cdots = 2 \cdot 2^n + 3 \cdot n$
...	...

In this list, we verifies finite samples to reach a conclusion $S_n = 2^{n+1} + 3 \cdot n$, which may not be reliable if we don't verity all cases. But it is impossible for us to do so when the number n goes to very big number even the infinity. From two examples above we notice that a conclusion about a conjecture from finite number of evidences is not reliable unless all evidences are true. Thus we have two types of induction.

1) **Incomplete induction** Incomplete induction reaches conclusion by verifying finite evidences of a conjecture. So it is not reliable when the number n goes beyond the number of evidences verified.

2) **Complete induction** Complete induction reaches conclusion by verifying all evidences of a conjecture. Then it is a reliable reasoning.

▶ Deductive Reasoning

In contrast, **deductive reasoning** is a reasoning to reach a conclusion for a particular conjecture using general principles (concept, axiom, and theorem). Unlike induction, deduction is not used to explorer new rule or theorem. It can be briefly described like

$$\boxed{general \rightarrow particular}$$

Basically deductive reasoning works in the following steps.
1) State a general principle.
2) Research features of a particular conjecture.
3) Reach a conclusion by applying the general principle to the conjecture.

For example, to prove the sequence $a_n = 2 \cdot 3^n$ is a geometric sequence, we use deductive reasoning as below.

- The definition of geometric sequences   (general principle)
- Research the sequence $a_n = 2 \cdot 3^n$ (particular conjecture)
- Because $a_{n+1}/a_n = 3$ is a constant number,   (applying general rule)
  the sequence $a_n = 2 \cdot 3^n$ is a geometric sequence.   (reach a conclusion)

▶ Mathematical Induction

> **Mathematical Induction**
> Mathematical induction is a reasoning combined induction and deduction, which is used to prove a mathematical statement (proposition) regarding natural numbers n.

Mathematical induction is a complete induction but most regular inductions are incomplete inductions. Although called mathematical induction, it actually applies theorem to prove individual case (*deduction*) and recursion to walk through all cases (*induction*). Therefore mathematical induction is a complete induction and reliable reasoning.

The statement to be proved by mathematical induction is about natural number n, which means there are n particular cases for that statement (*statement family*). We will walk through them one by one as natural number n increases. In other words, we will prove a series of statements not one statement.

In this book we will discuss the first principle of mathematical induction and its some variations as below.

- The first principle of mathematical induction (FPMI)
- The second principle of mathematical induction (SPMI)
- Mathematical induction by previous terms (MIPT)
- Mathematical induction by jumping (MIJ)

The first principle of mathematical induction (FPMI) is the fundamental of all variations. Let's take FPMI as example to explain how mathematical induction works. Let $P(n)$ represent a statement regarding a natural numbers n. If we prove the statement in regular inductive way we will prove it one by one when $n = 0, 1, 2, 3, \ldots$. But it is impossible for us to work in that way if natural number n goes a very huge number and even infinity. Now we introduce a recursive strategy to work on behalf of us. Briefly, if the truth of any statement, say $P(n+1)$, is derived from its previous one, say $P(n)$, all statements are true. Here the tricky is there must have a initial statement as starting point. By this recursive way mathematical induction works through all number n.

▶ Notations and Terminologies

1) $\mathbb{N}$ The set of natural number $\mathbb{N} = \{1, 2, 3, ...\}$

2) $\mathbb{Z}$ The set of integer $\mathbb{Z} = \{..., -2, -1, 0, 1, 2, 3, ...\}$

3) $\in$ An element of a set.

                              For example, the expression, $n \in \mathbb{N}$, means that n is an element of the set of natural number $\mathbb{N}$.

4) $\forall$ It means "for all ...".

                              For example, "$\forall n \geq 1$, $n \in \mathbb{N}$" means for all natural numbers n.

5) $\Longrightarrow$ It has meaning of "represent" or "be equivalent to".

6) $P(n)$ It denotes a statement, sentence, and formula involving a natural number n.

                              For example, "$P(9)$ is true." means that the statement denoted by $P(9)$ is true.

7) $P(n)$:   It means "Let $P(n)$ denote a statement, sentence or formula involving a natural number n."

                              For example, the expression

$$P(n): \quad 2^n > n^2 \quad (n \in \mathbb{N}, n \geq 5),$$

means "Let $P(n)$ denote the inequality, $2^n > n^2$."

The sentence "$P(n)$ is true $\forall n \geq 5$." is equivalent to "The inequality $2^n > n^2$ is true for all natural numbers that are equal to or great than 5."

8) **Start point n_0**   A fixed natural number, denoted n_0 ($n_0 \in \mathbb{N}$), from which a mathematical induction starts. Normally $n_0 = 1$ but it is not necessary, n_0 can be zero or any finite natural number.

1.2 The First Principle of Mathematical Induction

▶ **Explanation**

In order to explain **the first principle of mathematical induction** (**FPMI**), we will use a game of walking on a swamp as an example. Suppose that a man is planing to walk infinite distance on a swamp using two wood boards. He must find out a strategy to realize it.

The man believes that he can walk on a swamp if the following requirements are met.
1) There are two boards available.
2) A board is required for each of a man's step and no step is skipped.
3) The first board, called starting point, can be laid successfully.
4) A board can be laid if its previous board is laid for stepping on.

Based on the above requirements, the man works out his strategy as below (Figure 1.2.1).
(1) Lay the first board on the first step.
(2) Stand on the first board and lay a board for the second step.
(3) Step on the latest board laid if it is ready for standing on, lift the previous board and lay it for the next step.
(4) Repeat (3).

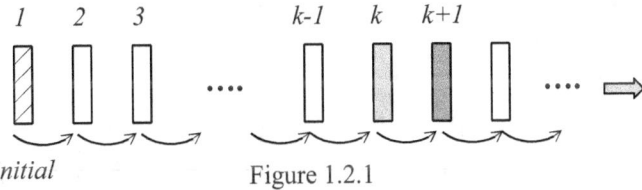

Figure 1.2.1

In this way the man can walk on a swamp all the way down. Someone may rise a question, "How is the man sure that he can lay a board infinitely?" The third requirement guarantees that any board can be laid if its previous one is laid. Thus the man is sure that if he stands on the latest board he can lay next board. The first principle of mathematical induction works in recursive way.

This game gives us a intuitive explanation about the first principle of mathematical induction. Let's translate the strategy above into mathematical language and you will understand how mathematical induction works easily.

Firstly we translate the following sentences in the game into mathematical language. To let reader refer to the steps in the strategy above we list step number, like (1) and (3), in the beginning of a sentence below.

-         The fact that the nth board was laid. ⟹ $P(n)$
- (1)   "The first board was laid." ⟹ $P(1)$ is true.
- (3)   " ... if it is ready for ... " ⟹ Assume $P(k)$ is true.
- (3)   "The board for the next step was laid." ⟹ $P(k+1)$ is true.
-         "The man can walk on a swamp." ⟹ $P(n)$ is true $\forall n \geq 1$.

Now the strategy the man works out can be translated into the format of mathematical induction like below.

1) <u>Defining a statement to be proved.</u>

    Let $P(n)$ ($n \in \mathbb{N}$) be the statement.

    "A man can walk an infinite distance on a swamp using two boards."

2) <u>Basis step ($n=1$)</u>

    Prove that $P(1)$ is true.

    "The first board is laid."

3) <u>Inductive Step ($n \geq 1$)</u>

    Set inductive hypothesis:

        Assume that $P(k)$ is true ($k \geq n_0, k \in \mathbb{N}$).
        "Assume that the kth board is laid."

    Show $P(k+1)$:

        Show that the truth of $P(k)$ implies that $P(k+1)$ is true.
        "Show that the $(k+1)^{th}$ board can be laid if the kth board is laid."

4) <u>Conclusion</u>:

    By the FPMI, $P(n)$ is true $\forall n \geq 1$. In other words, the statement

    "The man can walk an infinite distance on a swamp by two boards."

    is true.

1.2 The First Principle of Mathematical Induction

> **The First Principle of Mathematical Induction (FPMI)**
>
> Let $P(n)$ denote a statement involving a natural numbers n ($n \in \mathbb{N}$).
>
> If
> - $P(1)$ is true and
> - the assumption that $P(k)$ ($k \in \mathbb{N}$) is true implies that $P(k+1)$ is also true,
>
> then $P(n)$ is true for all natural numbers $\forall\, n \geq 1$.

The induction hypothesis, "$P(k)$ ($k \in \mathbb{N}$) is true.", must be used in the proof of $P(k+1)$. In this way we walks through all steps $n \geq 1$. When n steps forward, $P(k+1)$ relies on $P(k)$, in turns the step $P(k+2)$ relies on the step $P(k+1)$ and so on. See the figure below.

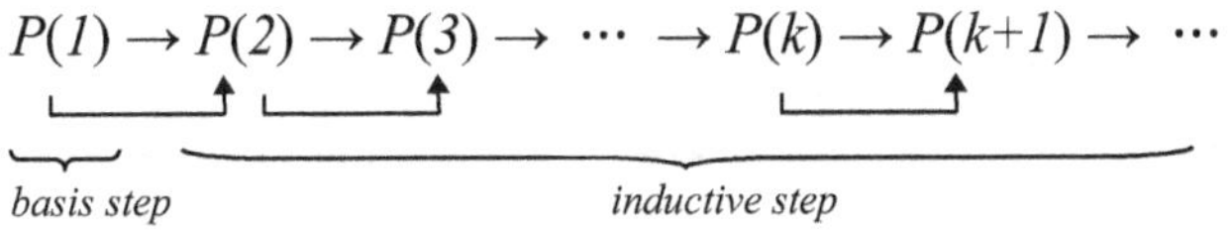

$$P(1) \to P(2) \to P(3) \to \cdots \to P(k) \to P(k+1) \to \cdots$$

$$\underbrace{}_{\text{basis step}} \quad \underbrace{}_{\text{inductive step}}$$

Figure 1.2.2

Note:
- You may say "You can not use a assumption to prove $P(k+1)$ in case it is not true." Yes, that's right. But the inductive hypothesis has its solid ground when it is the initial, $P(1)$. So the inductive hypothesis $P(k)$ is true when $k=1$.

- The truth of each step depends on its previous step and this significantly simplifies the proof for all natural number. In other words, the truth of each step can be proved from its previous step proved. As k in induction hypothesis is an arbitrary natural number, we can say it is true for all natural number.

- The truth of $P(k+1)$ is related to $P(k)$ only and does nothing with other previous terms, $P(1), P(2), \ldots, P(k-1)$.

▶ Extension of the First Principle

The first principle of mathematical induction (FPMI) is not restricted to start from 1 and it can start at 0 or any finite natural number n_0 ($n_0 \in \mathbb{N}$), called starting point. The FPMI starting $n_0 \neq 1$ is called **extension of the first principle (EFPMI)**. EFPMI works in the same way as FPMI except for starting point n_0.

Extension of the First Principle of Mathematical Induction

Let $P(n)$ denote a statement involving a natural numbers n ($n \in \mathbb{N}$).

If

- $P(n_0)$, where n_0 ($n_0 \in \mathbb{N}$) is a finite natural number, is true and
- the assumption that $P(k)$ ($k \geq n_0$) is true implies that $P(k+1)$ is also true,

then $P(n)$ is true $\forall\, n \geq n_0$.

▶ Proof Procedure

A recommended proof procedure for FPMI and EFPMI is given as below.

Defining $P(n)$

Let $P(n)$ be a statement, sentence, or formula involving natural number n.

Basis Step (n_0)

Prove that $P(n_0)$ is true.

Inductive Step ($n = k \geq n_0$)

- *Set inductive hypothesis*

    Assume $P(k)$ is true and this assumption must be used in next step.
- *Show $P(k+1)$*

    Arranging one side of the statement $P(k+1)$ to match the other side by the inductive hypothesis. If $P(k+1)$ is true, $P(n)$ is true $\forall\, n \geq n_0$.

Conclusion

Because $P(n)$ is true $\forall\, n \geq n_0$, the statement is true.

1.2 The First Principle of Mathematical Induction

Example 1.2.1 Sequence

If $\{a_n\}$ is a geometric sequence, then $a_n = a_1 \cdot r^{n-1}$ $(\forall n \geq 1)$.

Proof:

Defining $P(n)$

  Let $P(n)$ be the statement to be proven, i.e.

  $P(n)$: If $\{a_n\}$ is a geometric sequence, then $a_n = a_1 \cdot r^{n-1}$ $(\forall n \geq 1)$.

Basis Step $(n_0 = 1)$

  Look at the both sides of the formula, $a_n = a_1 \cdot r^{n-1}$ when $n = 1$,

  (L) a_1
  (R) $a_1 \cdot r^0 = a_1$.

  Because the both sides are equal, $P(1)$ is true.

Induction Step $(n = k \geq 1)$

- Set inductive hypothesis
  Assume that $P(k)$ ($k \in \mathbb{N}$, $k > 1$) is true, i.e.,
  $$a_k = a_1 \cdot r^{k-1}.$$

- Show $P(k+1)$
  Let's check both sides of the formula, $a_n = a_1 \cdot r^{n-1}$. When $n = k+1$, we have
  (L) the left side of the formula: a_{k+1}
       Using the inductive hypothesis above, it becomes
       $$a_{k+1} = a_k \cdot r = a_1 \cdot r^{k-1} \cdot r = a_1 \cdot r^{(k+1)-1}.$$
  (R) the right side of the formula is
       $$a_1 \cdot r^{(k+1)-1}$$

  Because (L) = (R), $P(k+1)$ is true.

Conclusion

  By the first principle of mathematical induction, $P(n)$ is true $\forall n \geq 1$. Thus the statement is true $\forall n \geq 1$.

Example 1.2.2 Sum of a Sequence

Prove that $1+2+2^2+\cdots+2^{n-1}=2^n-1$ ($\forall n \geq 1$, $n \in \mathbb{N}$)

Proof:

Defining $P(n)$

Let $P(n)$ be the statement to be proven, i.e.,
$$P(n): 1+2+2^2+\cdots+2^{n-1}=2^n-1 \quad (\forall n \geq 1).$$

Basis Step ($n_0 = 1$)

Check the both sides of $P(1)$,

(L) $2^{1-1}=1$
(R) $2^1-1=1$.

Because (L) = (R), $P(1)$ is true.

Induction Step ($n = k \geq 1$)

- Set inductive hypothesis

  Assume that $P(k)$ ($k \in \mathbb{N}$, $k > 1$) is true, i.e.,
  $$1+2+2^2+\cdots+2^{k-1}=2^k-1$$
  is true.

- Show $P(k+1)$

  Check both sides of $P(k+1)$.

  (L) $1+2+2^2+\cdots+2^{k-1}+2^k$

  Using the inductive hypothesis above, it becomes
  $$2^k-1+2^k=2^{k+1}-1.$$

  (R) $2^{k+1}-1$

  Because (L) = (R), $P(k+1)$ is true.

Conclusion

By the first principle of mathematical induction, $P(n)$ is true $\forall n \geq 1$. Thus the statement is true $\forall n \geq 1$.

1.2 The First Principle of Mathematical Induction

Example 1.2.3   Factor of a Formula

Prove that $x-y$ is a fact of $x^n - y^n$ ($x, y \in \mathbb{Z}$, $x \neq y$, $n \in \mathbb{N}$, $\forall n \geqslant 1$), where x and y are any integers.

Proof:

Defining $P(n)$

Let $P(n)$ be the statement to be proven, i.e.

$$P(n): x-y \text{ is a fact of } x^n - y^n.$$

Basis Step ($n_0 = 1$):

Because $x^1 - y^1 = x - y$, $x - y$ is a factor of $x^1 - y^1$, $P(1)$ is true.

Induction Step ($n = k \geqslant 1$):

- Set inductive hypothesis

   Assume that $P(k)$ ($k \in \mathbb{N}$, $k > 1$) is true. In other words, it is assumed that $x - y$ is a fact of $x^k - y^k$, i.e.,

   $$x^k - y^k = (x-y)i \ (i \in \mathbb{Z}).$$

- Show $P(k+1)$

   When $n = k+1$, we have

   $$x^{k+1} - y^{k+1} = (x-y)x^k + y(x^k - y^k)$$

   Look at the right side of the equation above. Because $x - y$ is a fact of the first term, $(x-y)x^k$, and by the induction hypothesis above, $x - y$ is a factor of the second term, $y(x^k - y^k)$, as well. Then $x - y$ is a fact of $x^{k+1} - y^{k+1}$ and $P(k+1)$ is true.

Conclusion

By the first principle of mathematical induction, $P(n)$ is true $\forall n \geqslant 1$. Thus the statement is true $\forall n \geqslant 1$.

Example 1.2.4   Divisibility

Prove that n^3+2n is divisible by 3 ($n \in \mathbb{N}$, $\forall n \geqslant 1$).

Proof:

Defining $P(n)$

Let $P(n)$ be the statement to be proven, i.e.,
$$P(n): n^3+2n \text{ is divisible by 3.}$$

Basis Step ($n_0 = 1$):

Because $1^3+2\cdot 1=3$, $P(1)$ is true.

Induction Step ($n=k \geqslant 1$):

- Set inductive hypothesis

  Assume $P(k)$ ($k \in \mathbb{N}$, $k>1$) is true. In other words, i.e., k^3+2k is divisible by 3.

  Let $$k^3+2k=3i \qquad (i \in \mathbb{Z}).$$

- Show $P(k+1)$

  When $n=k+1$, $P(k+1)$ becomes
  $$\begin{aligned}(k+1)^3+2(k+1) &= k^3+3k^2+3k+1+2k+2 \\ &= \underline{k^3+2k}+3(k^2+k+1) \\ &= 3i+3(k^2+k+1) \quad \text{(by the induction hypothesis)} \\ &= 3(i+k^2+k+1)\end{aligned}$$

  Because $(i+k^2+k+1)$ is an integer, $(k+1)^3+2(k+1)$ is divisible by 3. Then $P(k+1)$ is true.

Conclusion

By the first principle of mathematical induction, $P(n)$ is true $\forall n \geqslant 1$. Thus the statement is true $\forall n \geqslant 1$.

1.2 The First Principle of Mathematical Induction

Example 1.2.5   Divisibility
Prove that $(n+1)(n+2)$ is divisible by 2 ($\forall n \geq 1$, $n \in \mathbb{N}$).

Proof:

Defining $P(n)$

Let $P(n)$ be the statement to be proven, i.e.,

$$P(n): (n+1)(n+2) \text{ is divisible by } 2 \quad (\forall n \geq 1, n \in \mathbb{N}).$$

Basis Step ($n_0 = 1$):

Because $P(1): (1+1)(1+2) = 2$, $P(1)$ is true.

Induction Step ($n = k \geq 1$):

- Set inductive hypothesis

    Assume that $P(k)$ ($k \in \mathbb{N}$, $k > 1$) is true, i.e.,
    $$(n+1)(n+2) \text{ is divisible by } 2.$$
    Let $\quad (k+1)(k+2) = 2i \quad (i \in \mathbb{Z}).$

- Show $P(k+1)$

    When $n = k+1$, $P(k+1)$ becomes
    $$\begin{aligned}
    &((k+1)+1)((k+1)+2) \\
    &= (k+2)(k+3) \\
    &= k^2 + 3k + 2 + 2(k+1) \\
    &= (k+1)(k+2) + 2(k+1) \\
    &= 2i + 2(k+1) \quad \text{(by the induction hypothesis)} \\
    &= 2(i+k+1)
    \end{aligned}$$

    Because $(i+k+1)$ is an integer, $((k+1)+1)((k+1)+2)$ is divisible by 2. Then $P(k+1)$ is true.

Conclusion

By the first principle of mathematical induction, $P(n)$ is true $\forall n \geq 1$. Thus the statement is true $\forall n \geq 1$.

Example 1.2.6 Divisibility

Prove that $4^{2n+1}+3^{n+2}$ is divisible by 13 $\forall n \geq 1$. $(n \in \mathbb{N})$.

Proof:

Defining $P(n)$

Let $P(n)$ be the statement to be proven, i.e.,

$$P(n): 4^{2n+1}+3^{n+2} \text{ is divisible by } 13 \ \forall n \geq 1, \ (n \in \mathbb{N}).$$

Basis Step ($n_0 = 1$):

Because $P(1): 4^{2+1}+3^{1+2}=91$ is divisible by 13, $P(1)$ is true.

Induction Step ($n = k \geq 1$):

- Set inductive hypothesis

  Assume that $P(k): 4^{2k+1}+3^{k+2}$ is divisible by 13.

  Let $\qquad 4^{2k+1}+3^{k+2}=13 \cdot i \quad (i \in \mathbb{Z})$.

- Show $P(k+1)$

  $P(k+1)$ becomes

  $$\begin{aligned}
  & 4^{2(k+1)+1}+3^{(k+1)+2} \\
  &= 4^{2k+3}+3^{k+3} \\
  &= 16 \cdot 4^{2k+1}+3 \cdot 3^{k+2} \\
  &= 13 \cdot 4^{2k+1}+3 \cdot 4^{2k+1}+3 \cdot 3^{k+2} \\
  &= 13 \cdot 4^{2k+1}+3 \underbrace{(4^{2k+1}+3^{k+2})}_{} \quad \text{(by the induction hypothesis)} \\
  &= 13 \left(4^{2k+1}+3 \cdot i \right)
  \end{aligned}$$

  Because $4^{2k+1}+3 \cdot i$ is an integer, $4^{2(k+1)+1}+3^{(k+1)+2}$ is divisible by 13. Then $P(k+1)$ is true.

Conclusion

By the first principle of mathematical induction, $P(n)$ is true $\forall n \geq 1$. Thus the statement is true $\forall n \geq 1$.

1.2 The First Principle of Mathematical Induction

Example 1.2.7   *Identity*

Prove $1^3+2^3+3^3+\cdots+n^3=(1+2+3+\cdots+n)^2$ ($n\in\mathbb{N}, n\geqslant 1$).

Proof:

Defining $P(n)$

Let $P(n)$ be the identity to be proven, i.e.
$$P(n): 1^3+2^3+3^3+\cdots+n^3=(1+2+3+\cdots+n)^2.$$

Basis Step ($n_0 = 1$)

Check both sides of the identity $P(1)$ we have the followings.
(L) 1^3
(R) 1^2
Because (L) = (R), $P(1)$ is true.

Induction Step ($n = k \geqslant 1$)

- Set inductive hypothesis

  Assume that
  $$P(k): 1^3+2^3+3^3+\cdots+k^3=(1+2+3+\cdots+k)^2 \ (k\in\mathbb{N}, k>1)$$
  is true.

- Show $P(k+1)$

  (L)   The left side of the identity becomes
  $$1^3+2^3+\cdots+k^3+(k+1)^3=(1+2+3+\cdots+k)^2+(k+1)^3$$
  $$=(\frac{k(k+1)}{2})^2+(k+1)^3$$
  $$=\frac{k^2(k+1)^2+2^2(k+1)^3}{2^2}$$
  $$=\frac{k^4+6k^3+13k^2+12k+4}{2^2}$$
  $$=(\frac{(k+1)(k+2)}{2})^2$$
  $$=(1+2+3+\cdots+k+(k+1))^2$$

  (R)   The right side of the identity is
  $$(1+2+3+\cdots+k+(k+1))^2.$$
  Because (L) = (R), $P(k+1)$ is true.

Conclusion

By the first principle of mathematical induction, $P(n)$ is true $\forall n \geqslant 1$. Thus the identity is true $\forall n \geqslant 1$.

Example 1.2.8 Identity

Prove $\dfrac{1}{1\cdot 5}+\dfrac{1}{5\cdot 9}+\cdots+\dfrac{1}{(4n-3)(4n+1)}=\dfrac{n}{4n+1}$ ($n\in\mathbb{N}$, $\forall n\geqslant 1$).

Proof:

Defining $P(n)$

Let $P(n)$ be the identity to be proven, i.e.

$$P(n): \dfrac{1}{1\cdot 5}+\dfrac{1}{5\cdot 9}+\cdots+\dfrac{1}{(4n-3)(4n+1)}=\dfrac{n}{4n+1} \quad (n\in\mathbb{N}, \forall n\geqslant 1).$$

Basis Step ($n_0 = 1$)

Check both sides of the identity $P(1)$ we have the followings.

(L) $\quad \dfrac{1}{1\cdot 5}=\dfrac{1}{5}$

(R) $\quad \dfrac{1}{4\cdot 1 +1}=\dfrac{1}{5}$

Because both sides are equal, $P(1)$ is true.

Induction Step ($n = k \geqslant 1$)

- Set inductive hypothesis
  Assume that
  $$P(k): \dfrac{1}{1\cdot 5}+\cdots+\dfrac{1}{(4k-3)(4k+1)}=\dfrac{k}{4k+1} \quad (k\in\mathbb{N}, k>1)$$
  is true.

- Show $P(k+1)$
  (L) The left side of the identity becomes
  $$(\dfrac{1}{1\cdot 5}+\cdots+\dfrac{1}{(4k-3)(4k+1)})+\dfrac{1}{(4(k+1)-3)(4(k+1)+1)}$$
  $$=\dfrac{k}{4k+1}+\dfrac{1}{(4(k+1)-3)(4(k+1)+1)}$$
  $$=\dfrac{(4k+1)(k+1)}{(4k+1)(4k+5)}=\dfrac{(k+1)}{4(k+1)+1}.$$

  (R) The right side of the identity is $\dfrac{(k+1)}{4(k+1)+1}$.

  Because (L) = (R), $P(k+1)$ is true.

Conclusion

By the first principle of mathematical induction, $P(n)$ is true $\forall n \geqslant 1$. Thus the statement is true $\forall n \geqslant 1$.

1.2 The First Principle of Mathematical Induction

Example 1.2.9   Inequality

Prove that $x-y$ is a fact of $x^n - y^n$ ($x, y \in \mathbb{Z}$, $x \neq y$, $n \in \mathbb{N}$, $\forall n \geq 1$), where x and y are any integers.

Proof:

Defining $P(n)$

　Let $P(n)$ be the statement to be proven, i.e.
$$P(n): x-y \text{ is a fact of } x^n - y^n.$$
　We have starting point $n_0 = 1$ and length of step $l_s = 1$.

Basis Step ($n_0 = 1$)

　Because $x^1 - y^1 = x - y$, $x - y$ is a factor of $x^1 - y^1$. So $P(1)$ is true.

Induction Step ($n = k \geq 1$)

- *Set inductive hypothesis*
  Assume that $P(k)$ ($k \in \mathbb{N}$, $k > 1$) is true, i.e.,
  $$x^k - y^k = (x-y)i \quad (i \in \mathbb{Z}).$$

- *Show $P(k+1)$*
  When $n = k+1$, we have
  $$x^{k+1} - y^{k+1} = (x-y)x^k + y(x^k - y^k)$$

  From the right side of the equation above, $x - y$ is a fact of the first term, $(x-y)x^k$, and by the induction hypothesis, $x - y$ is a factor of the second term, $y(x^k - y^k)$, as well. Then $x - y$ is a fact of $x^{k+1} - y^{k+1}$ and $P(k+1)$ is true.

Conclusion

　By the first principle of mathematical induction, $P(n)$ is true $\forall n \geq 1$. Thus the statement is true $\forall n \geq 1$.

Example 1.2.10 Inequality

Prove $\dfrac{1}{2^1-1}+\dfrac{1}{2^2-1}+\cdots+\dfrac{1}{2^n-1}<2$ $(n\in\mathbb{N}, n\geq 1)$.

Proof:

Defining $P(n)$

Let $P(n)$ be the inequality to be proven, i.e.

$$P(n): \dfrac{1}{2^1-1}+\dfrac{1}{2^2-1}+\cdots+\dfrac{1}{2^n-1}<2 \quad (n\in\mathbb{N}, n\geq 1).$$

Basis Step ($n_0=1$)

Check both sides of the inequality $P(1)$ we have the followings.

(L) $\dfrac{1}{2^1-1}=1$

(R) 2

Because (L) < (R), the inequality holds when $n=1$ and $P(1)$ is true.

Induction Step ($n=k\geq 1$)

- Set inductive hypothesis
  Assume that $P(k)$: $\dfrac{1}{2^1-1}+\dfrac{1}{2^2-1}+\cdots+\dfrac{1}{2^k-1}<2$ $(k\in\mathbb{N}, k>1)$ is true.

- Show $P(k+1)$
  (L) The left side of the inequality becomes

  $$\dfrac{1}{2^1-1}+\dfrac{1}{2^2-1}+\cdots+\dfrac{1}{2^k-1}+\dfrac{1}{2^{k+1}-1}$$

  $$=\dfrac{1}{2^1-1}+\dfrac{1}{2}\left(\dfrac{1}{2^1-\frac{1}{2}}+\dfrac{1}{2^2-\frac{1}{2}}+\cdots+\dfrac{1}{2^k-\frac{1}{2}}\right) \quad (a)$$

  Since $\dfrac{1}{2^k-\frac{1}{2}}<\dfrac{1}{2^k-1}$, (a) $<\dfrac{1}{2^1-1}+\dfrac{1}{2}\left(\underline{\dfrac{1}{2^1-1}+\dfrac{1}{2^2-1}+\cdots+\dfrac{1}{2^k-1}}\right)$ (b)

  By the induction hypothesis above, (b) $<\dfrac{1}{2^1-1}+\dfrac{1}{2}\cdot 2=2$.

  (R) The right side of the inequality is 2.
  Because (L) < (R), $P(k+1)$ is true.

Conclusion

By the first principle of mathematical induction, $P(n)$ is true $\forall n\geq 1$. Thus the inequality is true $\forall n\geq 1$.

1.2 The First Principle of Mathematical Induction

Example 1.2.11   Inequality
Prove $|\sin nx| \leq n|\sin x|$ $(n \in \mathbb{N}, x \in \mathbb{R})$.

Proof:

Defining $P(n)$

Let $P(n)$ be the inequality to be proven, i.e.
$$P(n): |\sin nx| \leq n|\sin x| \quad (n \in \mathbb{N}, x \in \mathbb{R}).$$

Basis Step ($n_0 = 1$)

Check both sides of the inequality $P(1)$ we have the followings.
(L) $|\sin x|$
(R) $|\sin x|$
Because (L) = (R), $P(1)$ is true.

Induction Step ($n = k \geq 1$)

- Set inductive hypothesis
  Assume that $P(k): |\sin kx| \leq k|\sin x|$ $(k \in \mathbb{N}, k > 1)$ is true.
- Show $P(k+1)$
  (L)   The left side of the inequality becomes
  $$\begin{aligned}|\sin(k+1)x| &= |\sin(k+1)x| \\ &= |\sin(kx+x)| \\ &= |\sin kx \cos x + \sin x \cos kx| \\ &\leq |\sin kx| \cdot |\cos x| + |\sin x| \cdot |\cos kx| \\ &\leq |\sin kx| + |\sin x| \\ &\leq k|\sin x| + |\sin x| \\ &= (k+1)|\sin x|).\end{aligned}$$
  (R)   The right side of the inequality is $(k+1)|\sin x|$.
  Because (L) < (R), $P(k+1)$ is true.

Conclusion

By the first principle of mathematical induction, $P(n)$ is true $\forall n \geq 0$. Thus the inequality is true $\forall n \geq 0$.

Example 1.2.12 Inequality

Prove $1+2^2+3^3+\cdots+n^n<(n+1)^n$.

Proof:

Defining $P(n)$

    Let $P(n)$ be the inequality to be proven, i.e.
$$P(n): 1+2^2+3^3+\cdots+n^n<(n+1)^n \quad (n\in\mathbb{N}).$$

Basis Step ($n_0 = 1$)

    Check both sides of the inequality $P(1)$ we have the followings.

    (L)   1
    (R) $(1+1)^1 = 2$

    Because (L) < (R), $P(1)$ is true.

Induction Step ($n = k \geq 1$)

- Set inductive hypothesis

    Assume that
$$P(k): 1+2^2+3^3+\cdots+k^k<(k+1)^k \quad (k\in\mathbb{N}, k>1)$$
    is true.

- Show $P(k+1)$

    (L)   The left side of the inequality becomes
$$1+2^2+3^3+\cdots+k^k+(k+1)^{k+1}.$$
    By the induction hypothesis above,
$$1+2^2+3^3+\cdots+k^k+(k+1)^{k+1}<\underline{(k+1)^k}+(k+1)^{k+1}$$
$$=(k+1)^k(k+2)$$

    (R)   The right side of the inequality is
$$((k+1)+1)^{k+1}=(k+2)^k(k+2).$$

    Because $k > 1$ and $(k+1)^k<(k+2)^k$, $(k+1)^k(k+2)<(k+2)^k(k+2)$,

    Then (L) < (R) and $P(k+1)$ is true.

Conclusion

    By the first principle of mathematical induction, $P(n)$ is true $\forall n \geq 1$. Thus the inequality is true $\forall n \geq 1$.

1.2 The First Principle of Mathematical Induction

Example 1.2.13   *Inequality*

Prove $\dfrac{a^n+b^n}{2} \geq \left(\dfrac{a+b}{2}\right)^n$ $(a>0, b>0, n \in \mathbb{N})$.

Proof:

<u>Defining $P(n)$</u>
    Let $P(n)$ be the inequality to be proven, i.e.
$$P(n): \dfrac{a^n+b^n}{2} \geq \left(\dfrac{a+b}{2}\right)^n.$$

<u>Basis Step ($n_0 = 1$)</u>
    Check both sides of the inequality $P(1)$ we have the followings.
    (L) 1
    (R) $(1+1)^1 = 2$
    Because (L) < (R), $P(1)$ is true.

<u>Induction Step ($n = k \geq 1$)</u>
- Set inductive hypothesis

    Assume that $P(k)$: $\dfrac{a^k+b^k}{2} \geq \left(\dfrac{a+b}{2}\right)^k$ ($k \in \mathbb{N}$, $k > 1$) is true.

- Show $P(k+1)$

    (L)   The left side of the inequality becomes $\dfrac{a^{k+1}+b^{k+1}}{2}$.

    By the induction hypothesis above, $\dfrac{a^k+b^k}{2} \cdot \dfrac{a+b}{2} \geq \left(\dfrac{a+b}{2}\right)^{k+1}$.   (a)

    Because $(a-b)$ and $(a^k - b^k)$ have the same sign,
$$\dfrac{a^{k+1}+b^{k+1}}{2} - \dfrac{a^k+b^k}{2} \cdot \dfrac{a+b}{2} = \dfrac{(a-b)(a^k-b^k)}{4} \geq 0$$
$$\dfrac{a^{k+1}+b^{k+1}}{2} > \dfrac{a^k+b^k}{2} \cdot \dfrac{a+b}{2}.$$

    From (a), $\dfrac{a^{k+1}+b^{k+1}}{2} \geq \left(\dfrac{a+b}{2}\right)^{k+1}$.

    Thus $P(k+1)$ is true.

    (R)   The right side of the inequality is $\left(\dfrac{a+b}{2}\right)^{k+1}$.

    Then (L) $\geq$ (R) and $P(k+1)$ is true.

<u>Conclusion</u>
    By the first principle of mathematical induction, $P(n)$ is true $\forall n \geq 1$. Thus the inequality is true $\forall n \geq 1$.

Example 1.2.14 Natural Number

Prove that $\dfrac{n^5}{5}+\dfrac{n^3}{3}+\dfrac{7n}{15}$ ($n\in\mathbb{N}$) is a natural number.

Proof:

Defining $P(n)$

Let $P(n)$ be the statement to be proven, i.e.
$$P(n): \dfrac{n^5}{5}+\dfrac{n^3}{3}+\dfrac{7n}{15} \quad (n\in\mathbb{N})$$
is a natural number.

Basis Step ($n_0 = 1$)

Because $\dfrac{1^5}{5}+\dfrac{1^3}{3}+\dfrac{7\cdot 1}{15}=1$ is a natural number, $P(1)$ is true.

Induction Step ($n=k\geq 1$)

- *Set inductive hypothesis*

  Assume that
  $$P(k): \dfrac{k^5}{5}+\dfrac{k^3}{3}+\dfrac{7k}{15} \quad (k\in\mathbb{N},\ k>1)$$
  is a natural number.

- *Show $P(k+1)$*

  $P(k+1)$ becomes $\dfrac{(k+1)^5}{5}+\dfrac{(k+1)^3}{3}+\dfrac{7(k+1)}{15}$.

  $$= \underbrace{\left(\dfrac{k^5}{5}+\dfrac{k^3}{3}+\dfrac{7k}{15}\right)}_{P(k)} + \underbrace{(k^4+2k^3+3k^2+2k+1)}_{k\text{ is a natural number}}$$

  By the induction hypothesis above, the first part, $\dfrac{k^5}{5}+\dfrac{k^3}{3}+\dfrac{7k}{15}=P(k)$, is a natural number. Because k is a natural number, the second part,
  $$k^4+2k^3+3k^2+2k+1,$$
  is a natural number too. Then $P(k+1)$ is true.

Conclusion

By the first principle of mathematical induction, $P(n)$ is true $\forall n\geq 1$. Thus the inequality is true $\forall n\geq 1$.

1.2 The First Principle of Mathematical Induction

Example 1.2.15 *Inequality*

Prove $(1+a_1)(1+a_2)\cdots(1+a_n) \geq 1+a_1+a_2+\cdots+a_n$ if $a_1, a_2, \cdots, a_n \geq 0$.

Proof:

Defining $P(n)$

Let $P(n)$ be the inequality to be proven, i.e.

$$P(n): (1+a_1)(1+a_2)\cdots(1+a_n) \geq 1+a_1+a_2+\cdots+a_n \quad (\forall n \geq 1).$$

Basis Step ($n_0 = 1$)

Check both sides of the inequality and we have the followings.

(L) $1+a_1$
(R) $1+a_1$

Because (L) = (R), $P(1)$ is true.

Induction Step ($n = k \geq 1$)

- *Set inductive hypothesis*

  Assume that

  $$P(k): (1+a_1)(1+a_2)\cdots(1+a_k) \geq 1+a_1+a_2+\cdots+a_k \quad (k \in \mathbb{N}, k > 1)$$

  is true.

- *Show $P(k+1)$*

  (L) The left side of $P(k+1)$ is

  $$(1+a_1)(1+a_2)\cdots(1+a_k)(1+a_{k+1}) \quad (a)$$

  Expand (a) and we have

  $$\underbrace{(1+a_1)(1+a_2)\cdots(1+a_k)}_{\geq 1+a_1+a_2+\cdots+a_k} + \underbrace{(1+a_1)(1+a_2)\cdots(1+a_k)a_{k+1}}_{\geq 1} \quad (b)$$

  By the inductive hypothesis,

  $$(b) \geq 1+a_1+a_2+\cdots+a_k+a_{k+1}.$$

  (R) The left side of $P(k+1)$ is

  $$1+a_1+a_2+\cdots+a_k++a_{k+1}$$

  Thus $P(k+1)$ is true.

Conclusion

By the second principle of mathematical induction, $P(n)$ is $\forall n \geq 1$ true. Thus the statement is true $\forall n \geq 1$.

Example 1.2.16   The Binomial Theorem

Prove the binomial theorem

$$(x+y)^n = \sum_{m=0}^{n} {}_nC_m x^{n-m} y^m \quad (n, m \in \mathbb{N}, 0 \leq m \leq n).$$

Proof:

Defining P(n)

Let P(n) be the equation above, i.e.,

$$P(n): (x+y)^n = \sum_{m=0}^{n} {}_nC_m x^{n-m} y^m$$

Basis Step ($n_0 = 1$)

Check both sides of P(1) and we have the followings.
- (L) $(x+y)^1 = x+y$
- (R) ${}_1C_0 x + {}_1C_1 y = x+y$

Because (L) = (R), P(1) is true.

Induction Step ($n = k \geq 1$)

- Set inductive hypothesis
  Assume that P(k) is true, i.e.,

$$P(k): (x+y)^k = \sum_{m=0}^{k} {}_kC_m x^{k-m} y^m.$$

- Show P(k+1)

$$(x+y)^{k+1} = (x+y)^k (x+y)$$
$$= \left(\sum_{m=0}^{k} {}_kC_m x^{k-m} y^m \right)(x+y)$$

The $(m+1)^{th}$ term, the general term of P(k+1), becomes two parts.

$${}_kC_m x^{k-m} y^m \cdot (x+y) = {}_kC_m x^{k-m+1} y^m + {}_kC_m x^{k-m} y^{m+1} \quad \text{(a)}$$

Similarly we have the mth term and the $(m+2)^{th}$ term of P(k+1).

$${}_kC_{m-1} x^{k-m+1} y^{m-1} \cdot (x+y) = {}_kC_{m-1} x^{k-m+2} y^{m-1} + {}_kC_{m-1} x^{k-m+1} y^m \quad \text{(b)}$$
$${}_kC_{m+1} x^{k-m-1} y^{m+1} \cdot (x+y) = {}_kC_{m+1} x^{k-m} y^{m+1} + {}_kC_{m+1} x^{k-m-1} y^{m+1} \quad \text{(c)}$$

By the property of combinations we have

$$\begin{cases} {}_kC_{m-1} + {}_kC_m = {}_{k+1}C_m \\ {}_kC_m + {}_kC_{m+1} = {}_{k+1}C_{m+1} \end{cases}$$

1.2 The First Principle of Mathematical Induction

Then the first part of the $(m+1)^{th}$ term (a) will be combined with the second part of the mth term (b) to form the $(m+1)^{th}$ term of $(x + y)^{k+1}$. Similarly the second part of the $(m+1)^{th}$ term (a) and the first part of the $(m+2)^{th}$ term (c) are combined to form the $(m+2)^{th}$ term of $(x + y)^{k+1}$. In this way we find all rest terms of $(x + y)^{k+1}$. The figure below will give explanation more straight.

$$\begin{array}{ccc}
\text{the } m^{th} \text{ term of } (x+y)^k & \text{the } (m+1)^{th} \text{ term of } (x+y)^k & \text{the } (m+2)^{th} \text{ term of } (x+y)^k \\
\downarrow & \downarrow & \downarrow \\
\cdots + {}_kC_{m-1}x^{k-m+1}y^{m-1}\cdot(x+y) + & {}_kC_m x^{k-m}y^m\cdot(x+y) + & {}_kC_{m+1}x^{k-m-1}y^{m+1}\cdot(x+y) + \cdots
\end{array}$$

$$\cdots + {}_kC_{m-1}x^{k-m+2}y^m + {}_kC_{m-1}x^{k-m+1}y^{m+1} + {}_kC_m x^{k-m+1}y^m + {}_kC_m x^{k-m}y^{m+1} + {}_kC_{m+1}x^{k-m}y^{m+1} + \cdots$$

$$\cdots + ({}_kC_{m-1}+{}_kC_m)x^{k-m+1}y^m \quad + \quad ({}_kC_m+{}_kC_{m+1})x^{k-m}y^{m+1} + \cdots$$

$$\downarrow \qquad\qquad\qquad \downarrow$$

$$\cdots + \quad {}_{k+1}C_m x^{k+1-m}y^m \quad + \quad {}_{k+1}C_{m+1}x^{k-m}y^{m+1} \quad + \cdots$$

$$\downarrow \qquad\qquad\qquad \downarrow$$

$$\text{the } (m+1)^{th} \text{ term of } (x+y)^{k+1} \qquad \text{the } (m+2)^{th} \text{ term of } (x+y)^{k+1}$$

Figure 1.2.3

Then

$$(x+y)^{k+1} = {}_{k+1}C_0 x^{k+1} + {}_{k+1}C_1 x^k y + \cdots + {}_{k+1}C_m x^{k+1-m}y^m + \cdots + {}_{k+1}C_{k+1}y^{k+1}.$$

$$(x+y)^{k+1} = \sum_{m=0}^{k+1} {}_{k+1}C_m x^{k+1-m}y^m$$

Therefor $P(k+1)$ is true.

Conclusion

By the first principle of mathematical induction, $P(n)$ is true $\forall n \geq 1$. Thus the statement

$$(x+y)^n = \sum_{m=0}^{n} {}_nC_m x^{n-m}y^m \qquad (n, m \in \mathbb{N}, 0 \leq m \leq n).$$

is true $\forall n \in \mathbb{N}$.

Example 1.2.17 Inequality

Prove that $2^n > n^2$ ($n \in \mathbb{N}, n \geq 5$).

Proof:

Defining $P(n)$

Let $P(n)$ be the inequality to be proven, i.e.
$$P(n):\ 2^n > n^2 \quad (n \in \mathbb{N}, n \geq 5).$$

Basis Step ($n_0 = 5$)

Check both sides of the inequality $P(5)$ we have the followings.

(L) 2^5
(R) 5^2

Because $2^5 > 5^2$, $P(5)$ is true.

Induction Step ($n = k \geq 5$)

- Set inductive hypothesis

  Assume that $P(k):\ 2^k > k^2$ ($k \in \mathbb{N}, k > 5$) is true.

- Show $P(k+1)$

  (L) The left side of the inequality becomes
  $$2^{k+1} = 2 \cdot 2^k$$

  By the induction hypothesis above,
  $$2 \cdot 2^k > 2 \cdot k^2$$

  Since $k > 4$ and $k^2 > 4 \cdot k > 2k + 1$,
  $$2 \cdot k^2 = k^2 + k^2 > k^2 + 4 \cdot k > k^2 + 2k + 1.$$

  We have $2^{k+1} > k^2 + 2k + 1$.

  (R) The right side of the inequality is $(k+1)^2$.

  Because $k^2 + 2k + 1 = (k+1)^2$, (L) > (R). Then $P(k+1)$ is true.

Conclusion

By the first principle of mathematical induction, $P(n)$ is true $\forall n \geq 5$. Thus the inequality is true $\forall n \geq 5$.

1.2 The First Principle of Mathematical Induction

Example 1.2.18   Inequality

Prove $1+\dfrac{1}{2}+\dfrac{1}{3}+\cdots+\dfrac{1}{n}>\dfrac{2n}{n+1}$ ($n \in \mathbb{N}$, $n \geq 2$)

Proof:

Defining $P(n)$

Let $P(n)$ be the inequality, i.e., $P(n)$: $1+\dfrac{1}{2}+\dfrac{1}{3}+\cdots+\dfrac{1}{n}>\dfrac{2n}{n+1}$.

Basis Step ($n_0 = 2$)

- (L) $1+\dfrac{1}{2}=\dfrac{3}{2}$
- (R) $\dfrac{2\cdot 2}{2+1}=\dfrac{4}{3}$

Because (L) > (R), $P(2)$ is true.

Induction Step ($n = k \geq 2$)

- *Set inductive hypothesis*

  Assume that $P(k)$: $1+\dfrac{1}{2}+\dfrac{1}{3}+\cdots+\dfrac{1}{k}>\dfrac{2k}{k+1}$ ($k \in \mathbb{N}$, $k > 2$) is true.

- *Show $P(k+1)$*

  (L) The left side of the inequality becomes $1+\dfrac{1}{2}+\dfrac{1}{3}+\cdots+\dfrac{1}{k}+\dfrac{1}{k+1}$.

  By the induction hypothesis above,

  $$\underline{(1+\dfrac{1}{2}+\dfrac{1}{3}+\cdots+\dfrac{1}{k})}+\dfrac{1}{k+1}>\dfrac{2k}{k+1}+\dfrac{1}{k+1}=\dfrac{2k+1}{k+1}$$

  Let's compare the following two formulas

  $$\dfrac{2k+1}{k+1}=\dfrac{k}{k+1}+1, \quad \dfrac{2(k+1)}{(k+1)+1}=\dfrac{k}{k+2}+1.$$

  Because $k > 0$ and $\dfrac{k}{k+1}>\dfrac{k}{k+2}$, $\dfrac{2k+1}{k+1}>\dfrac{2(k+1)}{k+2}$. We have

  $$1+\dfrac{1}{2}+\dfrac{1}{3}+\cdots+\dfrac{1}{k}+\dfrac{1}{k+1}>\dfrac{2(k+1)}{(k+1)+1}.$$

  (R) The right side of the inequality is $\dfrac{2(k+1)}{(k+1)+1}$.

  Because (L) > (R), $P(k+1)$ is true.

Conclusion

By the first principle of mathematical induction, $P(n)$ is true $\forall n \geq 2$. Thus the inequality is true $\forall n \geq 2$.

Example 1.2.19 Identity

Prove $\cos x \cdot \cos 2x \cdot \cos 4x \cdots \cos 2^n x = \dfrac{\sin 2^{n+1} x}{2^{n+1} \sin x}$ ($n \in \mathbb{N}, n \geq 0$).

Proof:

Defining $P(n)$

Let $P(n)$ be the identity to be proven, i.e.

$$P(n): \cos x \cdot \cos 2x \cdot \cos 4x \cdots \cos 2^n x = \dfrac{\sin 2^{n+1} x}{2^{n+1} \sin x} \quad (n \in \mathbb{N}).$$

Basis Step ($n_0 = 0$)

Check both sides of the identity $P(0)$ we have the followings.

(L) $\cos x$

(R) $\dfrac{\sin 2x}{2 \sin x} = \cos x$

Because both sides are the same, $P(0)$ is true.

Induction Step ($n = k \geq 0$)

- Set inductive hypothesis
  Assume that
  $$P(k): \cos x \cdot \cos 2x \cdot \cos 4x \cdots \cos 2^k x = \dfrac{\sin 2^{k+1} x}{2^{k+1} \sin x} \quad (k \in \mathbb{N}, k > 0)$$
  is true.

- Show $P(k+1)$

  (L) $\cos x \cdot \cos 2x \cdot \cos 4x \cdots \cos 2^k x \cdot \cos 2^{k+1} x$

  $= \dfrac{\sin 2^{k+1} x}{2^{k+1} \sin x} \cdot \cos 2^{k+1} x$

  $= \dfrac{2 \sin 2^{k+1} x \cdot \cos 2^{k+1} x}{2^{k+2} \sin x}$

  $= \dfrac{\sin 2^{k+2} x}{2^{k+2} \sin x}$

  (R) $\dfrac{\sin 2^{k+2} x}{2^{k+2} \sin x}.$

  Because (L) = (R), $P(k+1)$ is true.

Conclusion

By the first principle of mathematical induction, $P(n)$ is true $\forall n \geq 0$. Thus the identity is true $\forall n \geq 0$.

1.2 The First Principle of Mathematical Induction

Example 1.2.20 *Geometry*

On a plane there are n $n \geq 2$ straight lines in such a way that no two of the lines are parallel and no three of the lines meet at a single point. Show there are $\frac{1}{2}n(n-1)$ intersections among of the lines.

Proof:

Defining $P(n)$

Let $P(n)$ be the statement, i.e.,

$P(n)$: There are $\frac{1}{2}n(n-1)$ intersections in total.

Basis Step ($n_0 = 2$)

Because there is one intersection between two lines and $\frac{1}{2}2(2-1)=1$, $P(2)$ is true.

Induction Step ($n = k \geq 2$)

- Set inductive hypothesis

  Assume that $P(k)$ is true, i.e., it is assumed there are $\frac{1}{2}k(k-1)$ intersections among k lines on a plane.

- Show $P(k+1)$

  $P(k+1)$ means there are $k+1$ lines on a plane. The $(k+1)^{th}$ line has k intersections with other k lines, so

  $$P(k+1) = P(k) + k$$
  $$= \frac{1}{2}k(k-1) + k$$
  $$= \frac{1}{2}(k+1)k$$
  $$= \frac{1}{2}(k+1)((k+1)-1)$$

  Then $P(k+1)$ is true.

Conclusion

By the second principle of mathematical induction, $P(n)$ is true $\forall n \geq 2$. Thus the statement is true $\forall n \geq 2$.

Example 1.2.21 *Geometry*

Prove that the sum of the interior angles of any convex polygon with n vertex is
$$f(n)=(n-2)\pi \quad (n\in \mathbb{N}, n\geq 3).$$
(Note: A polygon is called convex if any of its diagonals lies completely inside or on its boundary.)

Proof:

<u>Defining $P(n)$</u>

Let $P(n)$ be the statement to be proven, i.e.

$P(n)$: The sum of the interior angles of any convex polygon with n vertex is
$$f(n)=(n-2)\pi \quad (n\in \mathbb{N}, n\geq 3)$$

<u>Basis Step ($n_0 = 3$)</u>

It is a triangle when $n = 3$. Because the sum of the interior angles of a triangle is π and $f(3)=\pi$, $P(3)$ is true.

<u>Induction Step ($n=k\geq 3$)</u>

- *Set inductive hypothesis*

    Assume that $P(k)$: The sum of the interior angles of any convex polygon with k vertex, $V_1 V_2 \cdots V_k$, is
    $$f(k)=(k-2)\pi.$$

- *Show $P(k+1)$*

    When $n = k+1$, $P(k+1)$ becomes a convex polygon with $k+1$ vertex, $V_1 \cdots V_k V_{k+1}$. We can divide it into two parts, the triangle $\triangle V_1 V_k V_{k+1}$ and the polygon $V_1 V_2 \cdots V_k$, by drawing a line $V_1 V_k$. Obviously the sum of the interior angles of the polygon $V_1 \cdots V_k V_{k+1}$ becomes
    $$f(k+1)=f(k)+\pi.$$
    By the induction hypothesis above,
    $$f(k+1)=(k-2)\pi+\pi$$
    $$=((k+1)-2)+\pi$$
    Then $P(k+1)$ is true.

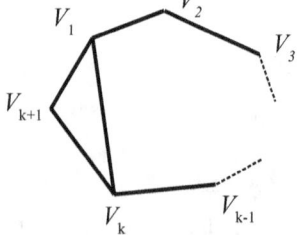

Figure 1.2.3

<u>Conclusion</u>

By the first principle of mathematical induction, $P(n)$ is true $\forall n \geq 3$. Thus the statement is true $\forall n \geq 3$.

1.2 The First Principle of Mathematical Induction

Example 1.2.22   Find and Prove a Statement

Given $1+3+5+\cdots+(2n-3)+(2n-1)+(2n-3)+\cdots+5+3+1$ $(n \in \mathbb{N})$
1) Find the sum of above expression when $n = 1, 2, 3$ and 5.
2) Find out the formula of the sum.
3) Prove that the formula is true for all natural number n.

Solution:

Let $S(n)$ be the sum of the formula given.

1) Find the sum when $n = 1, 2, 3$ and 5.

$S(1) = 1$
$S(2) = 4$
$S(3) = 9$
$S(5) = 25$

2) Find out the formula of the sum

Let $S(n) = 1+3+5+\cdots+(2n-3)+(2n-1)+(2n-3)+\cdots+5+3+1$.

Divide the above into three parts

$$[1+3+5+\cdots+(2n-3)]+(2n-1)+[(2n-3)+\cdots+5+3+1]$$
$$\underbrace{}_{1} \quad \underbrace{}_{2} \quad \underbrace{}_{3}$$

Add part 1 and 3 first and we have

$$[1+3+5+\cdots+(2n-3)]+[(2n-3)+\cdots+5+3+1]$$
$$= \underbrace{(2n-2)+\cdots+(2n-2)}_{n-1}$$
$$= (2n-2)(n-1).$$

Then $S(n)=(2n-2)(n-1)+(2n-1)$
$$= 2n^2 - 2n + 1$$

Now we reached the following statement

$$1+3+5+\cdots+(2n-3)+(2n-1)+(2n-3)+\cdots+5+3+1 = 2n^2-2n+1.$$

3) Prove that the statement above is true for all natural number n ($n \in \mathbb{N}$).

<u>Defining $P(n)$</u>

Let $P(n)$ be the statement to be proven, i.e.

$P(n)$: $1+3+\cdots+(2n-3)+(2n-1)+(2n-3)+\cdots+3+1 = 2n^2-2n+1.$

<u>Basis Step ($n_0 = 1$)</u>

Let's check both sides of $P(1)$ we have the followings.
(L) 1
(R) 1
Because (L) = (R), $P(1)$ is true.

Induction Step ($n=k\geq 1$)

- Set inductive hypothesis
  Assume that
  $P(k)$: $1+3+\cdots+(2k-3)+(2k-1)+(2k-3)+\cdots+3+1=2k^2-2k+1$
  ($k\in\mathbb{N}$, $k>1$) is a natural number.
- Show $P(k+1)$
  (L) The left side of $P(k+1)$ becomes
  $\underline{1+3+\cdots+(2k-3)+(2k-1)}+(2k+1)+(2k-1)+\underline{(2k-3)+\cdots+3+1}$ (a)
  By the induction hypothesis above,
  $$(a) = 2k^2-2k+1+(2k+1)+(2k-1)$$
  $$= 2k^2+2k+1$$
  (R) The right side of $P(k+1)$ is
  $$2(k+1)^2+2(k+1)+1$$
  $$= 2k^2+2k+1.$$
  Because (L) = (R), $P(k+1)$ is true.

Conclusion

By the first principle of mathematical induction, $P(n)$ is true $\forall n\geq 1$. Thus the formula is true $\forall n\geq 1$.

1.2 The First Principle of Mathematical Induction

Example 1.2.23   *Find and Prove a Statement*

Given a convex polygon with n vertices $n \in \mathbb{N}, n \geq 3$.
1) Find the number of diagonals $n = 3, 4, 5, 6, 7,$ and 8.
2) Figure out the formula of the number of diagonals
3) Prove that the formula is true $\forall\, n \geq 3$.

Solution:

1) Find the number of diagonals of the polygon with n vertices when $n = 3$ to 8.

We use induction to find the pattern of the relation between the number of diagonals of a polygon and the number of its vertices. Let $D(n)$ be the number of diagonals.

$$D(3) = 0 \text{ (triangle)}$$
$$D(4) = 2 \text{ (quadrilateral)}$$
$$D(5) = 5$$
$$D(6) = 9$$
$$D(7) = 14$$
$$D(8) = 20.$$

2) Figure out the formula of the number of diagonals of the polygon with n vertices

From the above, we are going to find the pattern by arranging them. Roughly, from $D(4)$, $D(6)$, and $D(8)$ we can guess that the formula wanted may include $\dfrac{n}{2}$.

$$D(4) = 2 = \frac{4}{2} = \frac{n}{2} \cdot 1$$

$$D(6) = 9 = \frac{n}{2} \cdot 3$$

$$D(8) = 20 = \frac{n}{2} \cdot 5$$

The formula could be

$$D(n) = \frac{n}{2} \cdot d(n). \qquad\qquad (a)$$

Further, after comparing (a) with $D(4)$, $D(6)$, and $D(8)$, we notice that $d(4) = 1$, $d(6) = 3$ and $d(8) = 5$ so we guess $d(n) = n - 3$. We have

$$D(n) = \frac{n}{2}(n-3) \qquad\qquad (b)$$

The formula (b) is validated by verifying it with $D(3)$, $D(5)$, and $D(7)$. As we known that the formula (b) validated by finite samples can not guarantee its validity. Even though it is verified for $n = 3, 4, 5, 6, 7,$ and 8, we are still not sure if it is validate for all natural number n. But this pattern recognition gives us a prototype to be proved by mathematical induction.

3) Prove that the statement above is true $\forall n \geq 3$.

   Defining P(n)

   Let $P(n)$ be the statement to be proven, i.e.

   $P(n)$: The number of the number of diagonals of a convex polygon with n vertices is

   $$f(n) = \frac{1}{2} n(n-3) \qquad (n \in \mathbb{N}, n \geq 3)$$

   Basis Step ($n_0 = 3$)

   The polygon is a triangle and there is no diagonal within it and $f(3) = 0$. Then $P(3)$ is true.

   Induction Step ($n = k \geq 3$)

   - Set inductive hypothesis

     Assume that $P(k)$: The number of diagonals of a convex polygon with k vertices is

     $$f(k) = \frac{1}{2} k(k-3) \qquad (k \in \mathbb{N}, k \geq 3)$$

   - Show $P(k+1)$

     When $n = k+1$, $P(k+1)$ becomes a convex polygon with $k+1$ vertices $V_1 \cdots V_{k+1}$ by adding a vertex V_{k+1} outside the polygon $V_1 \cdots V_k$ (see Figure 1.2.4). We get $k-2$ diagonals by connecting the vertex V_{k+1} to vertices $V_2 \cdots V_{k-1}$. As the side $V_1 V_k$ becomes a diagonal now, the total number of diagonals of the polygon $V_1 \cdots V_{k+1}$ is

     $$\begin{aligned} f(k+1) &= f(k) + (k-2) + 1 \\ &= \frac{1}{2} k(k-3) + (k-2) + 1 \\ &= \frac{1}{2} (k+1)((k+1)-3) \end{aligned}$$

     Then $P(k+1)$ is true.

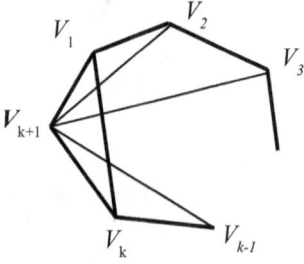

Figure 1.2.4

Conclusion

By the first principle of mathematical induction, $P(n)$ is true $\forall n \geq 3$. Thus the statement is true $\forall n \geq 3$.

2 Variations of Mathematical Induction

For different circumstances, there are several variations of the first principle of mathematical induction (FPMI). The first principle is used in such situation that the term $P(k+1)$ is derived from its previous term $P(k)$. But sometimes, $P(k+1)$ is derived from a particular, part or all of its previous terms. It is why we need variations of mathematical induction to deal with these circumstances. Here we show the following three variations used often.

1. **The Second Principle of Mathematical Induction (SPMI)**

    The second principle of mathematical induction, called **strong induction** or **complete induction**, has strong restriction on its induction hypothesis. It is assumed that all of the previous terms of the term $P(k+1)$ are true. SPMI is used in such case that the term $P(k+1)$ relies on all of its previous terms.

2. **Mathematical Induction by Previous Terms (MIPT)**

    Sometime the term $P(k+1)$ relies on a number of its previous terms. For this case, MIPT's induction hypothesis is loosen up and it is assumed that a number of the previous terms of the term $P(k+1)$ are true. For instance, MIPT can be used to prove a statement about a recursive sequence.

3. **Mathematical Induction by Jumping (MIJ)**

    In some cases, a term may be derived from a particular one of its previous terms, which two terms are separated by a gap. For example, the term $P(k+m)$ is derived from the term $P(k)$. It is like for the induction to jump m steps ahead from the term $P(k)$. For MIJ's induction hypothesis, it is assumed that the term $P(k)$ is true in order to show the term $P(k+m)$.

Among these four inductions, the second principle (SPMI) can replace mathematical induction by previous terms (MIPT) and the first principle (FPMI) because of SPMI's induction hypothesis strongly restricted.

2.1 The Second Principle of Mathematical Induction

▶ **Explanation**

In the first principle of mathematical induction (FPMI), we look back one step $P(k)$ to prove $P(k+1)$ but sometimes we need to look back more steps. That is where **the second principle of mathematical induction (SPMI)**, called **strong induction**, comes. Actually both FPMI and SPMI are equivalent except for their inductive hypothesis. We will use the game about walking on a swamp in Chapter 1 but the game is different now. This time the man in the game has as many boards as he needs. In order to transport these boards, he needs to build a track from the first board to the current board his standing on. To assure this track's success it is important that one board is laid on each of all previous steps.

The man believes that he can walk on a swamp if these requirements are met.
1) There are as many boards as needed.
2) A board is required for each step and no step is skipped.
3) The first board can be laid.
4) Standing on the the kth board, the man can lay a board for the $(k+1)^{th}$ step if all of the first k boards are laid.

Then the man works out his strategy as below (see Figure 2.1.1).
(1) Lay a board on the first step n_0.
(2) Stand on the first board and lay a board on the second step n_0+1.
(3) Step on the latest laid board k to lay a board for the next step $(k+1)$ if all of previous boards $(n_0, n_0+1, \ldots, k)$ laid.
(4) Increase k by 1, repeat (3).

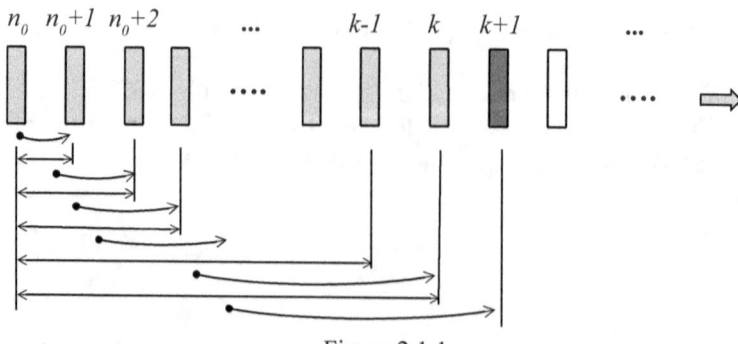

Figure 2.1.1

2.1 The Second Principle of Mathematical Induction

This strategy give us a clear picture about the second principle of mathematical induction. Now let's discuss it in mathematical language.

> **The Second Principle of Mathematical Induction (SPMI)**
>
> Let $P(n)$ denote a statement about natural numbers n ($n \in \mathbb{N}$).
>
> If
> - $P(n_0)$ ($n_0 \in \mathbb{N}$) is true, and
> - the assumption that $P(m)$ is true $\forall m$ ($n_0 \leq m \leq k$; $m, k \in \mathbb{N}$) implies that $P(k+1)$ is also true.
>
> then $P(n)$ is true $\forall n \geq n_0$.

Let's see how it works (see Figure 2.1.2).
1) Prove that the initial, $P(n_0)$, is true.
2) $P(n_0)$ leads to $P(n_0+1)$.
3) $P(n_0)$ and $P(n_0+1)$ lead to $P(n_0+2)$.
4) $P(n_0)$, $P(n_0+1)$, and $P(n_0+2)$ lead to $P(n_0+3)$.
5)   ......
6) $P(k-1)$, $P(k)$, $P(k+1)$ lead to $P(k+2)$.
7) $P(k)$, $P(k+1)$, $P(k+2)$ lead to $P(k+3)$.
8)   ......

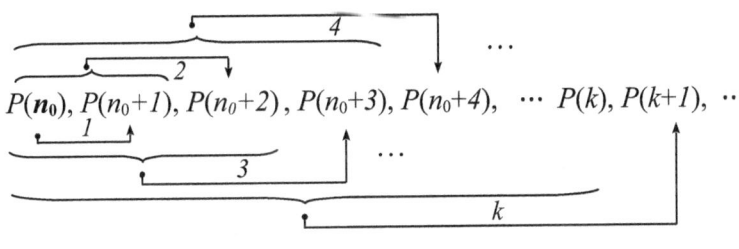

Figure 2.1.2

Note:
- There is one initial $P(n_0)$.
- The inductive hypothesis is different from that of FPMI and it is assumed that all of the first k terms, $P(n_0)$, $P(n_0+1)$, ..., $P(k)$, are true.
- To prove the term $P(k+1)$, you can use one, part, or all of its previous terms, $P(n_0)$, $P(n_0+1)$, ..., and $P(k)$.
- The inductive hypothesis must be used in the proof of $P(k+1)$.

▶ Proof Procedure

A recommended proof procedure for the second principle of mathematical induction (SPMI) is given as below.

Defining $P(n)$

   Let $P(n)$ be a statement to be proved.

Basis Step ($n_0=1$)

   Prove that $P(n_0)$ is true.

Induction Step ($n=k \geqslant n_0$)

- Set inductive hypothesis

   Assume that $P(m)$ is true $\forall\, m$ ($n_0 \leqslant m \leqslant k$), i.e., it is assumed that each of the first ($k-n_0+1$) terms $P(n_0), P(n_0+1), \ldots, P(k)$ are true.

- Show $P(k+1)$

   Prove $P(k+1)$ using the inductive hypothesis above

Conclusion

   If $P(k+1)$ is true, $P(n)$ is true $\forall\, n \geqslant n_0$. Thus the statement is true $\forall\, n \geqslant n_0$.

2.1 The Second Principle of Mathematical Induction

Example 2.1.1 Prime Numbers

Prove that any natural number n ($n \geq 2$) is prime number or a product of prime numbers.

Proof:

Defining $P(n)$

Let $P(n)$ be the statement

$P(n)$: Any positive integer n ($n \geq 2$) is prime number or a product of prime numbers.

Basis Step ($n_0 = 2$)

$P(2) = 2$. Because 2 is a prime $P(2)$ is true.

Induction Step ($n = k \geq 2$)

- *Set inductive hypothesis*

  Assume that $P(i)$ $2 \leq i < k$ is true, i.e., $P(2)$, $P(3)$, ..., $P(k)$ are true. In other words, it is assumed that each of the natural numbers 2, 3, ..., $k-1$, k is a prime number or a product of prime numbers

- *Show $P(k+1)$*

  There are two cases.
  - If $(k+1)$ is a prime number, $P(k+1)$ is true.
  - If $(k+1)$ is not a prime number, $P(k+1)$ can be factored as two natural numbers p and q like

  $$k+1 = p \cdot q.$$

  Because p and q are natural numbers which are neither 1 or k, we have

  $$\begin{cases} 1 < p < k \\ 1 < q < k \end{cases}$$

  By the induction hypothesis above, both $P(p)$ and $P(q)$ are true. It means that both are prime numbers or products of prime numbers.

  Thus $P(k+1)$ is true.

Conclusion

By the second principle of mathematical induction, $P(n)$ is true $\forall n \geq 2$. Thus the statement is true $\forall n \geq 2$.

Example 2.1.2   Prime Numbers

Prove $p_n < 2^{2^n}$ ($n \in \mathbb{N}$) if p_n is the nth prime, such as $p_1 = 2$ and $p_2 = 3$.

Proof:

Defining $P(n)$

Let $P(n)$ be the statement, i.e.,
$$P(n): \quad p_n < 2^{2^n} \quad (n \in \mathbb{N})$$

Basis Step ($n_0 = 1$)

Because $p_1 = 2$ and 2 is a prime, $P(1)$ is true.

Induction Step ($n = k \geq 1$)

- Set inductive hypothesis

  Assume that $P(m)$ $1 \leq m < k$ is true, i.e., $P(1)$, $P(2)$, ..., $P(k)$ are true. In other words, it is assumed that
  $$p_1 < 2^{2^1}, \, p_2 < 2^{2^2}, \, ..., \, p_k < 2^{2^k}.$$

- Show $P(k+1)$
  $$p_1 \cdot p_2 \cdots p_k < 2^{2^1} \cdot 2^{2^2} \cdot \cdots \cdot 2^{2^k} = 2^{2^1 + 2^2 + \cdots + 2^k}.$$

  Because both sides of the inequality above are integers, After adding *1* to the left side, it could be equal to the right side at most. So we have
  $$(p_1 \cdot p_2 \cdots p_k) + 1 \leq 2^{2^1 + 2^2 + \cdots + 2^k}$$
  $$= 2^{2^{k+1} - 2} < 2^{2^{k+1}}$$

  For any prime factor q of $(p_1 \cdot p_2 \cdots p_k) + 1$,
  $$q < 2^{2^{k+1}}$$

  Because these prime numbers, $p_1, p_2, \cdots, p_k$, are not prime factor of $(p_1 \cdot p_2 \cdots p_k) + 1$, we have
  $$p_k < q.$$
  $$p_{k+1} \leq q$$
  $$p_{k+1} < 2^{2^{k+1}}.$$

  Thus $P(k+1)$ is true.

Conclusion

By the second principle of mathematical induction, $P(n)$ is true $\forall n \geq 1$. Thus the statement is true $\forall n \geq 1$.

2.1 The Second Principle of Mathematical Induction

Example 2.1.3 Recursive Sequence

Prove $a_n < 2^n$ if $a_n = a_{n-1} + \frac{1}{2} a_{n-2}$ ($\forall n \geqslant 3$), $a_1 = 2$, and $a_2 = 3$.

Proof:

Defining $P(n)$

Let $P(n)$ denote the inequality to be proven, i.e.
$$P(n): a_n < 2^n \quad (\forall n \geqslant 3).$$

Basis Step ($n_0 = 3$)

Let's check both sides of the inequality, $P(3)$, we have the followings.

(L) $a_3 = a_2 + \frac{1}{2} \cdot a_1 = 3 + \frac{1}{2} \cdot 2 = 4$

(R) $2^3 = 8$

Because (L) < (R), $P(3)$ is true.

Induction Step ($n = k \geqslant 3$)

- *Set inductive hypothesis*

  Assume that $P(3), P(4), \ldots, P(k)$ are true, i.e.,
  $$a_3 < 2^3, a_4 < 2^4, \ldots, a_k < 2^k,$$
  are all true.

- *Show $P(k+1)$*

  (L) $\qquad a_{k+1} = a_k + \frac{1}{2} a_{k-1}.$

  By the induction hypothesis above, we know $a_k < 2^k$ and $a_{k-1} < 2^{k-1}$.

  Then $a_{k+1} = a_k + \frac{1}{2} \cdot a_{k-1} < 2^k + \frac{1}{2} \cdot 2^{k-1}$
  $$< (2^2 + 1) 2^{k-2}$$
  $$< 2^3 \cdot 2^{k-2} = 2^{k+1}$$

  (R) 2^{k+1}.

  Because $a_{k+1} < 2^{k+1}$, (L) < (R). Thus $P(k+1)$ is true.

Conclusion

By the second principle of mathematical induction, $P(n)$ is true $\forall n \geqslant 3$. Thus the inequality is true $\forall n \geqslant 3$.

Example 2.1.4   Recursive Sequence

If $a_n = a_{n-1} + a_{n-2}$ ($\forall n \geq 3$), $a_1 = 1$, and $a_2 = 1$, prove $a_n \geq 2 a_{n-2}$.

Solution:

Defining $P(n)$

Let $P(n)$ denote the inequality to be proven, i.e.
$$P(n): a_n \geq 2 a_{n-2} \quad (\forall n \geq 3).$$

Basis Step ($n_0 = 3$)

Let's check both sides of the inequality, $P(3)$, we have the followings.
(L) $a_3 = a_2 + a_1 = 2$
(R) $2 a_{3-2} = 2 a_1 = 22$
Because (L) = (R), $P(3)$ is true.

Induction Step ($n = k \geq 3$)

- Set inductive hypothesis
  Assume that $P(3), P(4), \ldots, P(k)$ are true, i.e., these inequalities,
  $$a_3 \geq 2 a_1, a_4 \geq 2 a_2, \ldots, a_k \geq 2 a_{k-2},$$
  are all true.

- Show $P(k+1)$
  (L) The left side of the inequality becomes
  $$a_{k+1} = a_k + a_{k-1}.$$
  By the induction hypothesis above, we know
  $$a_{k-1} \geq 2 a_{k-3}$$
  $$a_k \geq 2 a_{k-2}$$
  Then
  $$a_{k+1} = a_k + a_{k-1} \geq 2 \cdot a_{k-2} + 2 \cdot a_{k-3}$$
  $$a_{k+1} \geq 2 \cdot (a_{k-2} + a_{k-3})$$
  Since $a_{k-2} + a_{k-3} = a_{k-1}$, $a_{k+1} \geq 2 a_{k-1}$.
  (R) The right side of the inequality is $2 a_{k-1}$.
  Then (L) ≥ (R) and $P(k+1)$ is true.

Conclusion

By the second principle of mathematical induction, $P(n)$ is true $\forall n \geq 3$. Thus the inequality is true $\forall n \geq 3$.

2.1 The Second Principle of Mathematical Induction

Example 2.1.5   Recursive Sequence
Prove $a_n < 2^n + 3^n$ if $a_n = a_{n-1} + a_{n-2} + 2a_{n-3}$ ($\forall n \geq 4$), $a_1 = 1$, $a_2 = 2$, and $a_3 = 3$.

Solution:

<u>Defining P(n)</u>

Let $P(n)$ denote the inequality to be proven, i.e.
$$P(n): a_n < 2^n + 3^n \quad \forall n \geq 4.$$

<u>Basis Step ($n_0 = 4$)</u>

Let's check both sides of the inequality $P(4)$ we have the followings.
- (L) $a_4 = a_3 + a_2 + 2 \cdot a_1 = 7$
- (R) $2^4 + 3^4 = 97$

Because (L) < (R), $P(4)$ is true.

<u>Induction Step ($n = k \geq 4$)</u>

- *Set inductive hypothesis*

  Assume that $P(4), P(5), \ldots, P(k)$ are true, i.e., all these inequalities,
  $$a_4 < 2^4 + 3^4, \, a_5 < 2^5 + 3^5, \, \ldots, \, a_k < 2^k + 3^k,$$
  are true.

- *Show $P(k+1)$*

  (L) The left side of the inequality becomes
  $$a_{k+1} = a_k + a_{k-1} + 2a_{k-2}$$
  By the inductive hypothesis above, we have
  $$a_k < 2^k + 3^k, \, a_{k-1} < 2^{k-1} + 3^{k-1} \text{ and } a_{k-2} < 2^{k-2} + 3^{k-2}.$$
  Then $a_{k+1} = a_k + a_{k-1} + 2 \cdot a_{k-2}$
  $$< (2^k + 3^k) + (2^{k-1} + 3^{k-1}) + 2(2^{k-2} + 3^{k-2})$$
  $$< 2^{k+1} + 14 \cdot 3^{k-2}$$
  Since $14 < 3^3$ and $14 \cdot 3^{k-2} < 3^3 \cdot 3^{k-2} = 3^{k+1}$,
  $$a_{k+1} < 2^{k+1} + 3^{k+1}.$$

  (R) The right side of the inequality is $2^{k+1} + 3^{k+1}$.

  Then (L) < (R) and $P(k+1)$ is true.

<u>Conclusion</u>

By the second principle of mathematical induction, $P(n)$ is true $\forall n \geq 4$. Thus the inequality is true $\forall n \geq 4$.

2.2 Mathematical Induction by Previous Terms

▶ **Explanation**

In this section we are still going to use the game in Chapter 1 to explain **mathematical induction by previous terms (MIPT)**. This time the game will be changed and suppose that the man in the game has big foot so that he needs more boards to stand on at a time. Let m be the size of the man's foot and it means that m boards are needed for his standing on, where m is a finite natural number.

The man believes that he can walk on a swamp if the following requirements are met.

1) There are $m+1$ boards available.
2) A board is required for each step and no step is skipped.
3) There are m boards needed for a man to stand on at a time.
4) The first m boards can be laid successfully.
5) A board can be laid if its m previous boards have been laid for standing on.

The man works out his strategy as below (see Figure 2.2.1).

(1) Lay m boards for the first m steps, $n_0, n_0+1, \ldots, n_0+m-1$,
(2) Stand on the first m boards and lay a board on the $(n_0+m)^{th}$ step.
(3) Step on the latest m boards, i.e., the $(k-m+1)^{th}, \ldots,$ the $(k-1)^{th}$, and the kth steps, to lay a board on the $(k+m)^{th}$ step.
(4) Increase k by 1 and repeat (3).

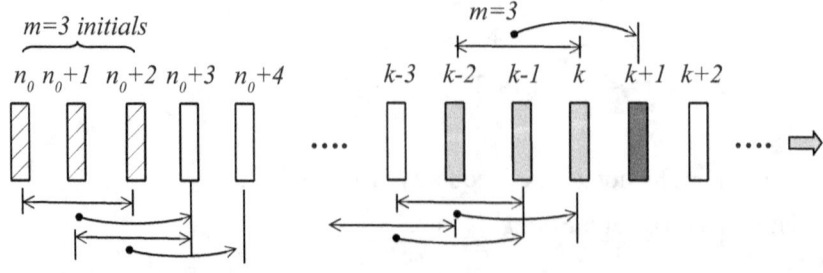

Figure 2.2.1

The strategy above explains mathematical induction by previous terms (MIPT) intuitively. Next we will explain it in mathematical language.

2.2 Mathematical Induction by Previous Terms

> **Mathematical Induction by Previous Terms (MIPT)**
>
> Let $P(n)$ denote a statement related to natural numbers n.
>
> If
> 1) $P(n_0), P(n_0+1), P(n_0+2), \cdots, P(n_0+m-1)$ $(m \in \mathbb{N})$ are true and
> 2) the assumption that $P(k), P(k+1), \ldots, P(k+m-1)$ are true implies $P(k+m)$ is also true.
>
> Then $P(n)$ is true $\forall\, n \geq n_0$.
>
> Where n_0 is a fixed natural number ($n_0 \in \mathbb{N}$) and m ($m \in \mathbb{N}$) is a finite natural number.

At first, let's take recursive sequences as example. We know that a term of a recursive sequence depends on its some previous terms. Similarly, for MIPT, a term $P(k+m)$ depends on its m previous terms.

For example, if $m = 3$, the term $P(k+3)$ is related to its three previous terms, $P(k)$, $P(k+1)$, and $P(k+2)$. Let's see how it works (see Figure 2.2.2).

1) Prove that the first 3 initials, $P(n_0), P(n_0+1)$, and $P(n_0+2)$ are true.
2) $P(n_0), P(n_0+1)$, and $P(n_0+2)$ lead to $P(n_0+3)$.
3) $P(n_0+1), P(n_0+2)$, and $P(n_0+3)$ lead to $P(n_0+4)$.
4) ⋯⋯
5) $P(k-1), P(k), P(k+1)$ lead to $P(k+2)$.
6) $P(k), P(k+1), P(k+2)$ lead to $P(k+3)$.
7) ⋯⋯

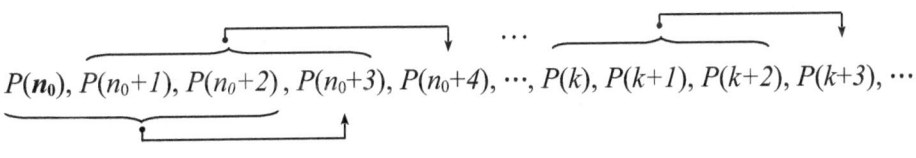

Figure 2.2.2

Note:
- There are m initials, $P(n_0), P(n_0+1), \cdots, P(n_0+m-1)$, are needed if the term $P(k+m)$ is related to its m previous terms. These m initials must be proved first.
- The inductive hypothesis must be used in proof of $P(k+m)$.

▶ Proof Procedure

A recommended proof procedure for mathematical induction by previous terms (MIPT) is given as below.

Defining $P(n)$
1) Let $P(n)$ be a statement to be proved.
2) Find out the number of the previous terms involved, denoted m, by the nature of the statement.

Basis Step ($n_0, n_0+1, \ldots, n_0+m-1$)
Prove that all m initials, $P(n_0)$, $P(n_0+1)$, ..., $P(n_0+m-1)$, are true.

Induction Step ($n = k \geq n_0$)
- *Set inductive hypothesis*
  Assume that $P(k-m+1)$, $P(k-m+2)$, $\cdots$, and $P(k)$ are true
- *Show $P(k+1)$*
  Prove $P(k+1)$ using the inductive hypothesis above

Conclusion
If $P(k+1)$ is true, $P(n)$ is true $\forall n \geq n_0$. Thus the statement is true $\forall n \geq n_0$.

2.2 Mathematical Induction by Previous Terms

Example 2.2.1   Recursive Sequence

If $a_1 = 1$, $a_2 = 1$, and $a_{n+2} = \frac{1}{2}(a_{n+1} + \frac{2}{a_n})$, prove $1 \leq a_n \leq 2 \ \forall n \geq 1$.

Solution:

Defining $P(n)$

    Let $P(n)$ denote the statement to be proven, i.e.
$$P(n): 1 \leq a_n \leq 2 \quad (\forall n \geq 1).$$

From the expression given, we know that the nth term, a_n, depends on its two previous terms, a_{n-1}, and a_{n-2}. That means that $P(n)$ depends on $P(n-1)$ and $P(n-2)$ and two initials, $P(1)$ and $P(2)$, are required.

Basis Step ($n_0 = 1, 2$)

    $P(1): a_1 = 1$, $P(1)$ is true.
    $P(2): a_2 = 1$, $P(2)$ is true.

Induction Step ($n = k \geq 1$)

- *Set inductive hypothesis*

  Assume that
  $$P(k): 1 \leq a_k \leq 2$$
  $$P(k+1): 1 \leq a_{k+1} \leq 2$$
  are all true.

- *Show $P(k+2)$*

  $P(k+2)$ is a_{k+2} and $a_{k+2} = \frac{1}{2}(a_{k+1} + \frac{2}{a_k})$.

  By the inductive hypothesis, we have
  $$\frac{1}{2} \leq \frac{a_{k+1}}{2} \leq 1$$
  $$\frac{1}{2} \leq \frac{1}{a_k} \leq 1.$$

  Then $1 \leq \frac{1}{2} a_{k+1} + \frac{1}{a_k} \leq 2.$

  Therefore $1 \leq a_{k+2} \leq 2$ and $P(k+2)$ is true.

Conclusion

    By the mathematical induction by previous terms, $P(n)$ is true $\forall n \geq 1$. Thus the statement is true $\forall n \geq 1$.

Example 2.2.2   Recursive Squence

If $a_{n+3} = a_{n+2} + 32 \cdot a_{n+1} - 60 \cdot a_n$ $(n \in \mathbb{N}, n \geq 1)$, $a_1 = 16$, $a_2 = 62$, and $a_3 = 274$, prove $a_n = 3 \cdot 2^n + 2 \cdot 5^n$.

Proof:

Defining P(n)

Let $P(n)$ denote the statement to be proven, i.e.

$$P(n): a_n = 3 \cdot 2^n + 2 \cdot 5^n \quad ((n \in \mathbb{N}, n \geq 1)).$$

In the expression given, the nth term, a_n, depends on its three previous terms, a_{n-1}, a_{n-2}, and a_{n-3}. It means that $P(n)$ depends on $P(n-1)$, $P(n-2)$ and $P(n-3)$ and three initials, $P(1)$, $P(2)$, and $P(3)$ are required.

Basis Step ($n_0 = 1, 2, 3$)

$P(1): a_1 = 16 = 3 \cdot 2^1 + 2 \cdot 5^1$ and $P(1)$ is true.
$P(2): a_2 = 62 = 3 \cdot 2^2 + 2 \cdot 5^2$ and $P(2)$ is true.
$P(3): a_3 = 274 = 3 \cdot 2^3 + 2 \cdot 5^3$ and $P(3)$ is true.

Induction Step ($n = k \geq 1$)

- Set inductive hypothesis

    Assume that $\quad P(k): a_k = 3 \cdot 2^k + 2 \cdot 5^k$
    $\quad\quad\quad\quad\quad\quad P(k+1): a_{k+1} = 3 \cdot 2^{k+1} + 2 \cdot 5^{k+1}$
    $\quad\quad\quad\quad\quad\quad P(k+2): a_{k+2} = 3 \cdot 2^{k+2} + 2 \cdot 5^{k+2}$
    are all true.

- Show $P(k+3)$
    (L) The left side of the statement is a_{k+3}.

    $a_{k+3} = a_{k+2} + 32 \cdot a_{k+1} - 60 \cdot a_k$
    $= (3 \cdot 2^{k+2} + 2 \cdot 5^{k+2}) + 32 \cdot (3 \cdot 2^{k+1} + 2 \cdot 5^{k+1}) - 60 \cdot (3 \cdot 2^k + 2 \cdot 5^k)$
    $= (3 \cdot 2^2 + 32 \cdot 3 \cdot 2 - 60 \cdot 3) \cdot 2^k + (2 \cdot 5^2 + 32 \cdot 2 \cdot 5 - 60 \cdot 2) \cdot 5^k$
    $= 24 \cdot 2^k + 250 \cdot 5^k$
    $= 3 \cdot 2^3 \cdot 2^k + 2 \cdot 5^3 \cdot 5^k$
    $= 3 \cdot 2^{k+3} + 2 \cdot 5^{k+3}$

    (R) The right side of the statement is $3 \cdot 2^{k+3} + 2 \cdot 5^{k+3}$.
    Then (L) = (R) and $P(k+3)$ is true.

Conclusion

By the mathematical induction by previous terms, $P(n)$ is true $\forall n \geq 1$. Thus the statement is true $\forall n \geq 1$.

2.3 Mathematical Induction by Jumping

▶ Explanation

To explain **mathematical induction by jumping** (**MIJ**) we will change the game in Chapter 1. Suppose that a man has arms measuring long as m, counting by steps, where m is a finite natural number. It means that he can exactly lay a board m steps ahead only,

The man believes that he can walk on a swamp if the following requirements are met.
1) There are $m+1$ boards available.
2) A board is required for each step and no step is skipped.
3) The first m boards can be laid successfully.
4) The man standing on the board on the kth step can exactly lay a board on the $(k+m)^{th}$ step only.

Based on the requirements above a strategy is worked out as below (see Figure 2.3.1).

(1) Lay m boards for the first m steps and let $k=1$.
(2) Step on the kth step and lay a board on the $(k+m)^{th}$ step.
(3) Let k increase by 1 and repeat (2).

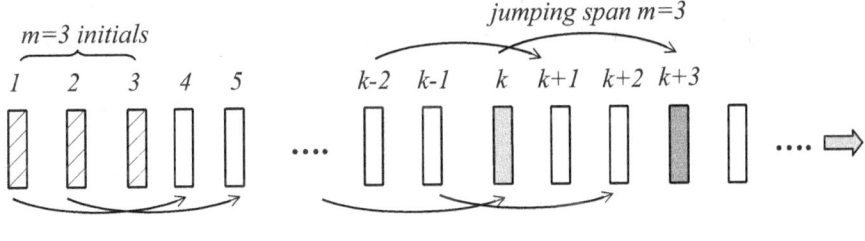

Figure 2.3.1

Using the game above we explain mathematical induction by jumping (MIJ) intuitively. Now we are going to discuss it in mathematical language.

Mathematical Induction by Jumping (MIJ)

Let $P(n)$ denote a statement related to a natural number n.

If

1) $P(n_0)$, $P(n_0+1)$, $\cdots$, $P(n_0+m-1)$ are true and
2) the assumption that $P(k)$ is true implies that $P(k+m)$ is also true.

then $P(n)$ is true $\forall\, n \geqslant n_0$.

Where n_0 and m, called **jumping span**, are fixed natural number $(n_0, m \in \mathbb{N})$.

Suppose jumping span $m=3$ and let's show how it works (see Figure 2.3.2).

1) Prove that the first 3 initials, $P(n_0)$, $P(n_0+1)$, and $P(n_0+2)$ are true.
2) $P(n_0)$ leads to $P(n_0+3)$.
3) $P(n_0+1)$ leads to $P(n_0+4)$.
4) $P(n_0+2)$ leads to $P(n_0+5)$.
5) ……
6) $P(k-1)$ leads to $P(k+2)$.
7) $P(k)$ leads to $P(k+3)$.
8) ……

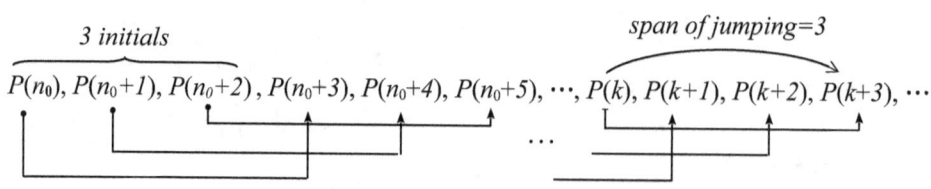

Figure 2.3.2

Note:
- There must be m initials to match with jumping span m. For example, if $m=3$, three initials $P(n_0)$, $P(n_0+1)$, and $P(n_0+2)$ should be proved first.
- $P(k+m)$ is derived from $P(k)$.
- The inductive hypothesis must be used in proof of $P(k+m)$.

2.3 Mathematical Induction by Jumping

▶ Proof Procedure

The following is a recommended proof procedure for mathematical induction by jumping (MIJ).

Defining $P(n)$

1) Let $P(n)$ be a statement to be proved.
2) Find out the jumping span of mathematical induction m, by the nature of the statement.

Basis Step ($n_0, n_0+1, \ldots, n_0+m-1$)

Prove that all m initials, $P(n_0), P(n_0+1), \ldots, P(n_0+m-1)$ are true.

Induction Step ($n = k \geq n_0$)

- *Set inductive hypothesis*
  Assume that $P(k)$ is true.
- *Show $P(k+m)$*
  Prove $P(k+m)$ using the inductive hypothesis above.

Conclusion

If $P(k+m)$ is true, $P(n)$ is true $\forall n \geq n_0$. Thus the statement is true $\forall n \geq n_0$.

Example 2.3.1   Amount of Postage

Prove that every amount of postage of 7 cents or more can be combined with 3-cent and 4-cent stamps.

Proof:

Defining $P(n)$

1) Let $P(n)$ be the statement

    "The amount of postage of n cents can be combined with 3-cent and 4-cent stamps $\forall\, n \geq 7$."

2) Decide span of jumping m

    - As 3-cent is the minimum amount to add to the accumulation of postage $P(n)$, the span of jumping $m = 3$.
    - Three initials are needed to match the span of jumping. Because 7-cents is the minimum amount of postage when combined with 3-cent and 4-cent, three initials are $P(7)$, $P(8)$, and $P(9)$.

Basis Step ($n_0 = 7, 8, 9$)

$P(7) = 7¢$ (one 3-cent and one 4-cent stamps)
$P(8) = 8¢$ (two 4-cent stamps)
$P(9) = 9¢$ (three 3-cent stamps)

Then $P(7)$, $P(8)$, and $P(9)$ are true.

Induction Step ($n = k \geq 7$)

- Set inductive hypothesis

    Assume that $P(k)$ ($k \geq n_0, k \in \mathbb{N}$) are true, i.e., the amount of postage of k-cent can be combined with 3-cent and 4-cent stamps.

- Show $P(k+3)$

    $P(k+3)$ represents $k+3$ cents and it can be written as
    $$P(k+3) = (k+3)¢$$
    $$= k¢ + 3¢.$$
    By the inductive hypothesis, $P(k) = \underline{k¢}$ is true then
    $$P(k+3) = \underline{P(k)} + 3¢.$$
    Thus $P(k+3)$ is true.

Conclusion

By the second principle of mathematical induction, $P(n)$ is true $\forall\, n \geq 7$. Thus the statement is true $\forall\, n \geq 7$.

2.3 Mathematical Induction by Jumping

Example 2.3.2 *Geometry*

Prove that a square can be subdivided into any number of smaller squares $n \geq 6$.

Proof:

Defining $P(n)$

1) Let $P(n)$ denote the statement, i.e.,

   $P(n)$: A square can be subdivided into any number of smaller squares $n \geq 6$.

2) Find out initials and span of jumping:

   - Because any square can be easily subdivided into four smaller squares, we can get three more squares for each subdivision. Then the span of jumping is 3.
   - Three initials are required to match the span of jumping and they are $P(6)$, $P(7)$, and $P(8)$.

Basis Step ($n_0 = 6, 7, 8$)

We can subdivide a square into 6, 7, and 8 smaller squares as below.

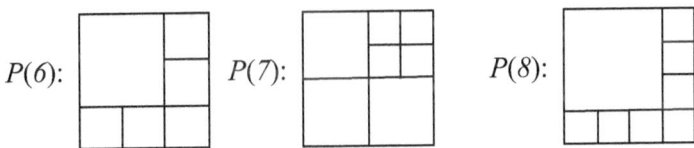

Figure 2.3.3

Then $P(6)$, $P(7)$, and $P(8)$ are true.

Induction Step ($n = k \geq 6$)

- Set inductive hypothesis

  Assume that $P(k)$ ($k \geq 6$) is true. That is assumed that a square can be subdivided into k smaller squares.

- Show $P(k+3)$

  Because any square within the square $P(k)$ can be easily subdivided four smaller squares, we can obtain $k + 3$ squares. That is that $P(k+3)$ is true.

Conclusion

By mathematical induction by jumping, the statement $P(n)$ is true $\forall n \geq 6$. In other words, the statement is true $\forall n \geq 6$.

399

Margin trade · 109, 357
margin trading · 59, 109, 110, 123, 362
Matcha · 181, 182, 185, 362
Merkle tree · 357
metadata · 357
MetaMask · 22, 26, 34, 42, 55, 57, 61, 115, 125, 131, 157, 184, 218, 222, 225, 227, 249, 290, 307, 309, 313, 316, 319, 335, 358, 366
mining · 358
MKR · 28, 29
Mooniswap · 265
MSD · 28

N

Nexus Mutual · 300
NFT · 343
non-custodian · 13
Non-Fungible Token · 343

O

Oasis · 29, 30, 31, 81, 181, 196
Oasis app · 31
offline wallet · 356
OPEN LONG POSITION · 115
OpenZeppeli · 304
options trading · 5, 303, 329
Opyn · 304, 305, 312, 314, 321, 329, 365
order books · 147, 150
oToken · 321

P

P2P lending · 294
P2P- Transactions · 13
PAX · 158
permissionless · 14
perpetual contract · 121, 122
perpetual contracts · 121, 122
Perpetual Trading · 126
phishing attack · 290
PICKLE · 222, 225, 226, 227, 228
Pickle Finance · 213, 222
Pickle Jar Swap · 223
POA Network · 147
Polkadot · 13, 263, 264, 379
Polkaswap · 263, 264, 373, 374
PoolTogether · 282
private pool · 232
procedural risks · 289
proof of work · 358
protocol risk · 290
put options · 303, 304, 305, 334

R

Radar Relay · 137, 362
rebalance · 274
REN · 158, 364

S

sBTC · 154, 156, 158
sBTC) · 154

Security Token Offering · 358

sETH · 156, 157, 211

sHEGIC · 340

slippage · 134, 158, 211

smart contract · 122, 127, 211, 232, 233, 357

smart contracts · 3, 11, 13, 41, 60, 127, 147, 174, 288, 300, 304, 312, 358

Smart Savings · 99, 101

SNX · 154, 211, 252, 253, 254, 255, 295, 364

Solidity · 3, 358

Sora · 4, 260, 263, 264, 268, 269, 270, 370, 372, 374, 376, 377, 378, 379

Sora Farm · 4, 260, 263, 264, 268

SORA v2 network · 269

Soramitsu · 264, 269

Soranomics · 269

spot trading · 59, 109, 122, 362

stablecoin · 28, 361

Stablecoin · 38

supply chain management · 3

sUSD · 156, 157, 158, 160, 253, 254, 255, 295, 349, 352, 353, 354, 355

SUSHI · 183, 186, 364

Swap · 129, 131, 150, 151, 152, 174, 176, 186, 193, 194, 195, 196, 212, 247, 248, 362

Swap OTC · 174, 176

Swaprate · 296, 298

SWERVE APP · 209

Swerve Finance · 208

SWRV · 208

Synthetix · 154, 156, 211

Synthetix Exchange · 252

T

technical risks · 288

Time Index · 292

Token · 57, 127, 133, 147, 157, 171, 186, 232, 358, 364

tokenized assets · 15

tokenized physical assets · 15

Tokenlon · 186, 189, 190, 362

TokenSets · 271

Torque · 4, 57

transactions · 13, 14, 15, 237, 282, 358

trustless · 11, 154

U

UDSC · 88

undercollateralized · 21, 41

UNI · 21, 195, 196, 222, 224, 225, 256, 305, 364

Uniswap · 4, 34, 81, 105, 127, 128, 156, 157, 181, 196, 211, 212, 213, 215, 222, 224, 225, 228, 238, 253, 256, 263, 265, 266, 267, 268, 269, 271, 275, 281, 301, 306, 321, 322, 324, 325, 326, 327, 328, 347, 349, 352, 363, 373

USDC · 21, 47, 53, 58, 59, 61, 62, 63, 67, 68, 69, 91, 93, 104, 109, 110, 111, 112, 122, 125, 126, 129, 131, 134, 135, 144, 208, 224, 235, 236, 237, 238, 239, 261,

279, 286, 295, 296, 299, 306, 309, 312, 313, 318, 347, 373

USDT · 21, 43, 47, 53, 55, 57, 125, 126, 161, 165, 176, 178, 183, 189, 190, 191, 195, 199, 204, 205, 208, 210, 259, 279, 286, 295, 373

Utilization Index · 292

V

VAL · 263, 265, 268, 269, 270, 373, 378, 379, 380

W

wallet address · 23, 128, 155, 168, 197, 271

WBTC · 21, 47, 232, 233, 295, 305, 330, 331, 332, 334, 335, 336, 338, 340, 364

WETH · 201, 202, 232, 233, 234, 235, 237, 305, 306

writeETH · 335, 336, 337

writeWBTC · 335, 336

Wyre · 35

X

XOR · 263, 265, 266, 268, 269, 270, 371, 372, 373, 374, 375, 376, 377, 378, 379, 380

Y

YAM · 4

Yearn Finance · 212, 361

yield farming · 11, 65, 136, 162, 210, 211, 212, 213, 222, 256, 257, 259, 260, 263, 271, 289, 358

Yield Farming · 4, 11, 210, 211, 359

Z

Zapper · 271, 272, 282, 361

Zerion · 162, 271, 283, 286

ZHEGIC · 341

ZRX · 21, 364

www.ingramcontent.com/pod-product-compliance
Lightning Source LLC
Chambersburg PA
CBHW070615220526
45466CB00001B/8